Mos

Reed-Solomon Related Codes

Mostafa El-Khamy

Reed-Solomon Related Codes

New Approaches to Performance Analysis and Decoding Algorithms

VDM Verlag Dr. Müller

Impressum/Imprint (nur für Deutschland/ only for Germany)

Bibliografische Information der Deutschen Nationalbibliothek: Die Deutsche Nationalbibliothek verzeichnet diese Publikation in der Deutschen Nationalbibliografie; detaillierte bibliografische Daten sind im Internet über http://dnb.d-nb.de abrufbar.

Alle in diesem Buch genannten Marken und Produktnamen unterliegen warenzeichen-, marken- oder patentrechtlichem Schutz bzw. sind Warenzeichen oder eingetragene Warenzeichen der jeweiligen Inhaber. Die Wiedergabe von Marken, Produktnamen, Gebrauchsnamen, Handelsnamen, Warenbezeichnungen u.s.w. in diesem Werk berechtigt auch ohne besondere Kennzeichnung nicht zu der Annahme, dass solche Namen im Sinne der Warenzeichen- und Markenschutzgesetzgebung als frei zu betrachten wären und daher von jedermann benutzt werden dürften.

Coverbild: www.purestockx.com

Verlag: VDM Verlag Dr. Müller Aktiengesellschaft & Co. KG
Dudweiler Landstr. 99, 66123 Saarbrücken, Deutschland
Telefon +49 681 9100-698, Telefax +49 681 9100-988, Email: info@vdm-verlag.de
Zugl.: Pasadena, California Institute of Technology, Ph.D. Thesis, 2006

Herstellung in Deutschland:
Schaltungsdienst Lange o.H.G., Berlin
Books on Demand GmbH, Norderstedt
Reha GmbH, Saarbrücken
Amazon Distribution GmbH, Leipzig
ISBN: 978-3-639-14948-7

Imprint (only for USA, GB)

Bibliographic information published by the Deutsche Nationalbibliothek: The Deutsche Nationalbibliothek lists this publication in the Deutsche Nationalbibliografie; detailed bibliographic data are available in the Internet at http://dnb.d-nb.de.

Any brand names and product names mentioned in this book are subject to trademark, brand or patent protection and are trademarks or registered trademarks of their respective holders. The use of brand names, product names, common names, trade names, product descriptions etc. even without a particular marking in this works is in no way to be construed to mean that such names may be regarded as unrestricted in respect of trademark and brand protection legislation and could thus be used by anyone.

Cover image: www.purestockx.com

Publisher:
VDM Verlag Dr. Müller Aktiengesellschaft & Co. KG
Dudweiler Landstr. 99, 66123 Saarbrücken, Germany
Phone +49 681 9100-698, Fax +49 681 9100-988, Email: info@vdm-publishing.com
Pasadena, California Institute of Technology, Ph.D. Thesis, 2006

Printed in the U.S.A.
Printed in the U.K. by (see last page)
ISBN: 978-3-639-14948-7

To My Family

&

To Peace.

Peace cannot be kept by force.
It can only be achieved by understanding.

—Albert Einstein

Preface

This work was inspired by Sudan's breakthrough that demonstrated that Reed-Solomon codes can correct more errors than previously thought. This breakthrough can render the current state-of-the-art Reed-Solomon decoders obsolete. Much of the importance of Reed-Solomon codes stems from their ubiquity and utility. This work takes a few steps toward a deeper understanding of Reed-Solomon codes as well as toward the design of efficient algorithms for decoding them.

After studying the binary images of Reed-Solomon codes, we proceeded to analyze their performance under optimum decoding. Moreover, we investigated the performance of Reed-Solomon codes in network scenarios when the code is shared by many users or applications. We proved that Reed-Solomon codes have many more desirable properties. Algebraic soft decoding of Reed-Solomon codes is a class of algorithms that was stirred by Sudan's breakthrough. We developed a mathematical model for algebraic soft decoding. By designing Reed-Solomon decoding algorithms, we showed that algebraic soft decoding can indeed approach the ultimate performance limits of Reed-Solomon codes. We then shifted our attention to products of Reed-Solomon codes. We analyzed the performance of linear product codes in general and Reed-Solomon product codes in particular. Motivated by these results we designed a number of algorithms, based on Sudan's breakthrough, for decoding Reed-Solomon product codes. Lastly, we tackled the problem of analyzing the performance of sphere decoding of lattice codes and linear codes, e.g., Reed-Solomon codes, with an eye on the tradeoff between performance and complexity.

Acknowledgements

It is a pleasure to take this opportunity to thank all the people who have touched my life and helped the dream of this work come true. I consider myself very fortunate to have Prof. Robert J. McEliece as my advisor. It is his brilliant lectures on information theory and the theory of error-correcting codes that made me love this field. His sharp insight, consistent guidance, constant encouragement, contagious enthusiasm, and friendly advice are all echoed throughout this work. His intelligent questions led to many of the results in this work. For all the things I have learned from him, I will always be indebted to him.

I am grateful to the members of my candidacy and defense committees, Prof. Robert J. McEliece, Prof. P. P. Vaidyanathan, Prof. Babak Hassibi, Prof. Steven Low at the California Institute of Technology, Prof. Dariush Divsalar at the Jet Propulsion Laboratory and Prof. Marc Fossorier at the University of Hawaii. I would also like to thank them for the unmatched help and generous support that they have provided to me and for their invaluable advice and constructive feedback.

The intellectual and stimulating environment at the California Institute of Technology had a huge impact on the quality of research presented in this work. I would like to thank Prof. Michelle Effros for hosting me in her research group during my master's year. I would like to thank many of those whom I had technical discussions with and those whom I have collaborated with on numerous research problems. I would like to thank Haris Vikalo for the engaging discussions we had, Roberto Garello for his patient advice, Makiko Kan for her careful feedback, Yuval Cassuto for his enthusiasm, Farzad Parvaresh for the fun and fruitful time we had while he was visiting our research group and Alex Vardy for his insightful comments. Without a doubt, the friendly environment created by the other students in my research group, over the past four years, was a key factor in making this work. I am grateful to Cedric Florens, Ravi Palanki, Jeremy Thorpe, Jonathan Harel, Edwin Seodormadji and Sarah Fogal for making my experience at Caltech such a wonderful one. My sincere thanks also go to my office-mates, Masoud Sharif, Mihailo Stojnic, Amir Farajidana, Radhika Gowaiker, Tareq Al-Naffouri, Chaitanya Rao, Weiyu Xu, Ali Vakili, Sormeh Shadbakht and Frederique Oggier for the enriching and pleasant atmosphere they have created.

My thanks also go to our friendly administrative assistants Shirley Betty and Linda Dozsa for their professional aid in all the administrative issues. Many thanks to Greg Fletcher at the Caltech-Y and Jim Endrizzi at the International Student Programs for all the social activities they have organized to make our stay at Caltech beneficial in so many ways.

Many thanks also go to my friends in the Teaching Assistant room at Alexandria University for the mutual encouragement we gave to each other. My thanks also go to those professors at Alexandria University who gave their best to see this happen.

This work has been made possible by the generous support of the National Science Foundation, Qualcomm Corp., Sony Corp. and the Lee Center for Advanced Networking.

My heartfelt thanks go to my parents, Said and Sanaa, and my sisters, Rasha, Rehab and Donia, with great appreciation and respect. Their generous love, extraordinary care and unconditional support has been with me all the way. I owe them so much, more than I can ever pay back, for always being there for me.

Thanks to God for making this dream unfold into reality.

Contents

List of Figures

Chapter 1

Introduction

The road to success is always under construction.

—Lily Tomlin

The now ubiquitous Reed-Solomon codes were invented in 1960 [93]. It was not until the late sixties when Berlekamp and Massey invented an efficient algorithm for decoding them [12]. Today, billions of dollars are invested in products, which carry error-correcting encoders and decoders, and millions of error-correcting codes are being decoded each minute. It is no exaggeration to say that at least three-quarters of the codes used today are Reed-Solomon codes. Reed-Solomon codes have many properties, such as their random-error-correction capability, burst-error-correction capability, and erasure-recovery capability, which make them very appealing for many applications. Their success can be attributed to the efficient encoding and decoding algorithms and their state-of-the-art integrated circuit implementations.

Everyone who has ever used a computer has in fact used a Reed-Solomon code. For decades Reed-Solomon codes have been used in the magnetic storage devices such as hard disks. With other breakthroughs in channel coding such as the invention of Turbo codes [13] and the resurrection of LDPC codes [45, 78] one might wonder if this is still the case. These codes, however, suffer from error-floor problems. If such codes were to be implemented for their capacity-approaching capability, Reed-Solomon (RS) codes will still be used as outer codes to cure their error-floor problems. Other storage devices such as compact discs (CDs) and digital versatile discs (DVDs) also standardize concatenated RS codes and RS product codes as their error-correcting codes. It is worth noting that storage devices are now making their way in our everyday devices such as cell phones, play stations, personal digital assistants (PDAs), digital music players, digital cameras and high-definition televisions. As we are in the trend of digitizing everything, we are in more need than ever for reliable storage space. Moreover, we need to be able to access this digital information quickly which translates to the need of having efficient decoding algorithms and high speed decoding circuits.

Without Reed-Solomon codes, deep space exploration might have simply been a dream. Reed-Solomon codes were used to encode the digital pictures sent to us by the Voyager space probe. Reed-Solomon is currently deployed in all probes in operation and will still be used in future missions. Reed-Solomon codes, concatenated with convolutional codes, have been the state-of-the-art channel codes for deep space communication. The 2004 Mars Exploration Rover mission that successfully sent two rovers Spirit and Opportunity to explore the Martian surface and geology had Reed-Solomon codes in operation. Similar standards of Reed-Solomon codes and concatenated Reed-Solomon codes are also used in satellite communication for digital video broadcasting.

Reed-Solomon codes have also been adopted as outer codes in the third generation (3G) wireless standard, CDMA2000 high-rate broadcast packet data air interface [1], and are expected to be used as outer codes in concatenated coding schemes for future fourth generation wireless systems. Hybrid automatic repeat request (H-ARQ) error control systems for asymmetric digital subscriber line (ADSL) access networks deploy block interleaved Reed-Solomon codes to maintain a high throughput and reliability. Interleaved Reed-Solomon codes are also the standard in high speed optical fiber networks operating at 10 Gbps. Amusingly, mailing services, such as the United States Postal Service (USPS), deploy a black-ink bar code, called PostBar, which is printed on packages for automatic mail sorting. PostBar uses a Reed-Solomon coding technique for error correction in case it is defected from mishandling the mail.

Almost forty years after the invention of the Berlekamp-Massey algorithm, we were surprised to realize that polynomial-time decoding algorithms can correct more errors in Reed-Solomon codes than previously thought. This breakthrough came with the invention of the Sudan [102] and Guruswami-Sudan [49] list-decoding algorithms for

RS codes, for which Sudan was awarded the prestigious Nevannlina prize. Rather than returning one codeword, list-decoding algorithms return a list of codewords. Although the concept of list decoding dates back to 1957 [39], it was not until 1997 [102] that we were able to efficiently list decode RS codes beyond their classical error-correction capability.

1.1 Contributions

Most of the research in this work was motivated and inspired by the theoretical break-through of the Guruswami-Sudan algorithm. Our first goal was to study the ultimate performance limits of Reed-Solomon codes. With the new advances in networking and the progress in ad hoc networking techniques, it was natural to think of RS codes as the code of choice in multiuser environments. This motivated us to study the performance of RS codes in multiuser settings. The Guruswami-Sudan algorithm did not make full use of the soft information at the channel output. Koetter and Vardy built on the Guruswami-Sudan algorithm and devised a soft-decision list-decoding algorithm for RS codes. This motivated us to study the ultimate performance of such soft-decision list-decoding algorithms. We designed soft-decision list-decoding algorithms for Reed-Solomon that perform better than previously known algorithms. In fact, the performance of our iterative list-decoding algorithm approaches the performance lim-its of RS codes at a reasonable complexity. As we see from the discussion above, RS product codes and concatenated RS are widely deployed in many applications. This motivated us to study the performance of linear product codes in general and RS prod-uct codes in particular. The performance limits of RS product codes showed that there is much room for improvement over the current decoding algorithms. This motivated us to study list-decoding of RS product codes. We designed and analyzed algebraic list-decoding algorithms for decoding RS product codes. We believe that such decod-ing algorithms can dramatically improve the performance of the widely deployed RS product codes. The Guruswami-Sudan algorithm can also be viewed as sphere decod-ing algorithm. A sphere decoder is one which will return a list of codewords within a certain sphere without actually searching all such codewords. Sphere decoders are currently the state of the art decoders in multiple input-multiple output (MIMO) wireless systems and have received a lot of attention. This connection to the Guruswami-Sudan algorithm motivated us to study the performance of sphere decoding of linear block codes in general and Reed-Solomon related codes in particular under various settings.

1.2 Outline

Next we give a more detailed outline of the contents and contributions of this work. This work is designed such that each chapter can be read separately. However, we do refer the reader to the results in other chapters whenever needed.

Chapter 2: Binary images of Reed-Solomon Codes [29, 28]:

Although there was a significant amount of research dedicated to developing better decoding algorithms for Reed-Solomon codes, there was little known about their fundamental operating limits and researchers relied on comparing the performance of their algorithms with other algorithms. Reed-Solomon codes are often defined over finite fields of characteristic two. In many applications, it is the binary image of the RS code that is transmitted over the channel. Whereas knowledge of the weight enumerator of a linear code is essential to analyze its performance, the binary weight enumerators of binary images of RS codes depend on the basis used to represent the symbols as bits. An averaged binary weight enumerator for RS codes is derived and is shown to closely estimate an exact one for a specific basis representation. Moreover, it has been shown that as the code length and the finite field size tend to infinity, the weight enumerator of the ensemble of binary images of Reed-Solomon codes approach that of a random code with the same dimensions.

By considering the performance of the ensemble of binary images of an RS code, rather than a specific binary image, we are able to develop tight upper bounds on the performance of the optimum maximum-likelihood decoder. We analyze both cases of soft-decision and hard-decision maximum-likelihood decoding. Observing that a code's performance at high signal-to-noise ratios relies heavily on its minimum distance, we analyzed the minimum distance of the binary image of a RS code. It is then shown that the ensemble of binary images of RS codes is asymptotically good.

Chapter 3: The Multiuser Error Probability of Reed-Solomon Codes [28, 32]:

Maximum distance separable (MDS) codes have many attractive properties which make them the code of choice in network scenarios and distributed coding schemes. Reed-Solomon codes are the most popular MDS codes. Given an arbitrary partition of the coordinates of a code, we introduce the partition weight enumerator which enumerates the codewords with a certain weight profile in the partitions. A closed form formula of the partition weight enumerator of maximum distance separable codes is derived. Using this result, some properties of MDS codes are discussed. In particular, we show that all coordinates have the same weight within the subcodes of constant weight codewords. The results are extended to the ensemble of binary images of MDS codes defined over finite fields of characteristic two. The error probability of Reed-Solomon codes in multiuser networks is then studied. This analysis can be extended to many network scenarios. For example, we analyze the case when a Reed-Solomon code (or its binary image) is shared among different users or applications. Such a system is likely to exist in wireless multiuser networks where the sensor nodes, of limited power, can communicate with a local base station in an error free manner. The local base station will then group their data symbols and encode them into a single codeword for transmission over a noisy channel to another cluster of nodes. After being decoded by the receiving base station, the multiuser data symbols are then routed to their desired destination.

Chapter 4: Algebraic Soft-Decision Decoding of Reed-Solomon Codes: Interpolation Multiplicity Assignments [31, 34]:

Decoding Reed-Solomon codes beyond half-the-minimum distance of the code is a major breakthrough in modern coding theory that was introduced by Sudan and Guruswami. After decades of bounded minimum distance decoding, the Guruswami-Sudan algorithm shows us how major achievements can be obtained by tackling hard problems in a different way. Moreover, this algorithm led to the pioneering work of Koetter and Vardy on algebraic soft-decision decoding. Some questions were posed to us.

What is the potential limit of algebraic soft decoding? Are there better algebraic soft-decision decoding algorithms? In an attempt to answer these questions we developed a mathematical framework for algebraic soft-decision decoding. We devised a new method, based on the Chernoff bound, for assigning interpolation multiplicities for algebraic soft-decision list decoding. We formulated the problem as a constrained optimization problem aiming at directly minimizing the decoder error probability. An iterative algorithm was devised for assigning the interpolation multiplicities for any desired interpolation cost. We were able to show that the potential performance of algebraic soft-decision decoding is much better than previously thought.

Chapter 5: Iterative Algebraic Soft-Decision Decoding of Reed-Solomon Codes [30, 33]:

We present an iterative soft-decision list-decoding algorithm for Reed-Solomon codes offering both complexity and performance advantages over previously known decoding algorithms. Our algorithm is a list-decoding algorithm which combines two powerful soft-decision decoding techniques which were previously regarded in the literature as competitive, namely, the Koetter-Vardy algebraic soft-decision decoding algorithm and belief propagation based on adaptive parity check matrices, recently proposed by Jiang and Narayanan. Building on the Jiang-Narayanan algorithm, we present a belief-propagation based algorithm with a significant reduction in computational complexity. We introduce the concept of using a belief-propagation based decoder to enhance the soft-input information prior to list decoding with an algebraic soft-decision decoder. Instead of assuming that all the received symbols are independent, we enhance the reliability of the received symbols based on the information about the code. We show that in such a setting algebraic soft-decision decoding can achieve near maximum-likelihood decoding with reasonable interpolation costs. Our algorithm can also be viewed as an interpolation multiplicity assignment scheme for algebraic soft-decision decoding of Reed-Solomon codes.

Chapter 6: Performance Analysis of Linear Product Codes [26, 27]:

Product RS codes are widely used, especially in data storage systems and digital video broadcast systems. The recent breakthroughs in decoding RS codes motivated us to investigate turbo decoding of RS product codes by iteratively decoding the component codes using algebraic soft-decision decoding. This led us to the natural question: What are the performance limits of linear product codes? It turned out that the weight enumerator of most linear product codes, and thus their maximum-likelihood performance, is very hard to determine. The analytical performance evaluation of product codes relied on the truncated union bound, which provides a low error rate approximation based on the minimum distance term only.

We approached the problem differently by introducing concatenated representations of product codes and applying them to compute the complete average enumerators of arbitrary product codes over an arbitrary finite field. The derivation of the weight enumerator of the product codes required the knowledge of the split weight enumerator of the component codes. We were able to derive simple closed form formulas of the split weight enumerator of some popular linear codes. Together with some of the results in the previous chapters, we were able to derive tight upper bounds on the soft-decision and hard-decision maximum-likelihood performance of linear product codes in general and Reed-Solomon product codes in particular. The weight enumerator of the ensemble of binary images of product Reed-Solomon codes were also derived. Our results show that Reed-Solomon product codes can have a performance very close to the capacity of the channel and that, unlike LDPC and Turbo codes, they do not seem to suffer from error floors. Our results predict the importance of devising low complexity efficient algorithms for decoding product codes.

Chapter 7: Algebraic List Decoding of Reed-Solomon Product Codes [84]:

The product code of two Reed-Solomon codes can be regarded as an evaluation code of bivariate polynomials, whose degrees in each variable are bounded. We propose to decode these codes with a generalization of the Guruswami-Sudan interpolation-based list-decoding algorithm. We devised a polynomial time list-decoding algorithm for two-dimensional Reed-Solomon product codes based on trivariate polynomial interpolation. It has a relative decoding radius of $(1 - \sqrt[6]{4R_p})$, where R_p is the rate of the product code. We also devise a generalized algorithm for decoding M-dimensional product codes with a relative decoding radius of $1 - {}^{M(M+1)}\!\!\sqrt{M^M R_p}$. We also propose another algorithm based on the observation that Reed-Solomon product codes are subcodes of Reed-Muller codes. We then deploy the Pellikaan-Wu interpretation of decoding Reed-Muller codes as subcodes of generalized Reed-Solomon codes to decode Reed-Solomon product codes. This algorithm is capable of correcting more errors as its relative decoding radius is $1 - \sqrt[4]{4R_p}$ for two-dimensional RS product codes and $1 - \sqrt[2M]{M^M R_p}$ for M-dimensional product codes.

Chapter 8: Performance Analysis of Sphere Decoders [35, 36, 37]: Sphere decoding algorithms are often used in wireless channels for decoding lattice codes and for detection in multiple antenna wireless systems. A sphere decoder is a decoder that will return the closest lattice point, if it exists within a specified search radius, without actually searching all lattice points. This directly connected to the Guruswami-Sudan algorithm which is a polynomial time algorithm with an asymptotic Hamming decoding radius that can be larger than half-the-minimum distance of the code. A large number of researchers focused on analyzing the complexity of soft-decision sphere decoders and developing algorithms with lower complexities. However, little research has been devoted to the performance analysis of sphere decoders. This motivated us to study the performance of sphere decoders and derive tight upper bounds on their performance under various settings. We considered both soft-decision and hard-decision sphere decoders. We also analyzed the performance on different channels and modulation schemes. To extend this analysis to sphere decoders that decode Reed-Solomon codes on the symbol level, such as the Guruswami-Sudan algorithm, we analyzed the performance of hard-decision sphere decoder on q-ary symmetric channels. For the sake of this analysis, we derived a tight upper bound on the performance of maximum-likelihood decoding of a linear code defined over a finite field of size q when transmitted over a q-ary symmetric channel. Our analysis of the performance of sphere decoders enable one to choose the decoding radius that best fits the desired performance, throughput and complexity of the system.

Chapter 2

Binary Images of Reed-Solomon Codes

Without the capacity to provide its own information, the mind drifts into randomness.

——Mihaly Csikszentmihalyi

Reed-Solomon (RS) codes are the most popular maximum distance separable (MDS) codes. For any linear (n, k, d) code (of length n, dimension k and minimum distance d) over any field, maximum distance separable (MDS) codes have the maximum possible minimum distance $d = n - k + 1$ [74]. MDS codes have many other desirable properties which made them the code of choice in many communication systems. MDS codes have the property that any k codeword coordinates can be considered as the information symbols in a systematic codeword and any k coordinates can be used to recover the information symbols. Moreover, punctured MDS codes are also MDS codes. Such properties made MDS codes a natural choice in Automatic-Repeat-Request (ARQ) communication systems (c.f., [116]). MDS codes are also used in the design of multi-cast network codes [122].

Maximum-likelihood (ML) decoding of linear codes, in general, and RS codes, in particular, is NP-hard [10, 50]. It remains an open problem to find polynomial-time decoding algorithms with near ML performance. The Guruswami-Sudan (GS) algorithm was the first polynomial time hard-decision decoding algorithm for Reed-Solomon

codes capable of correcting beyond half-the-minimum distance of the code at all rates
[49]. Moreover, the invention of the GS algorithm has spurred a significant amount
of research aiming at better soft-decision decoding algorithms for Reed-Solomon codes
(c.f., [76, 72, 31, 33, 65]).

Suppose a Reed-Solomon (RS) code is defined over a finite field of characteristic
two, then it is a common practice to send its binary image over the channel. In fact,
the binary image has a large burst-error-correction capability which is one of the main
reasons behind the ubiquitous use of RS codes. The decoder can either be a bit-level
decoder, which decodes the RS code as a binary code, or a symbol level decoder,
which treats the received word as a vector in the finite field. It is often the case that
hard-decision decoders, which do not make use of the reliability information from the
channel, are symbol based decoders. Such hard-decision decoders, as the Berlekamp-
Massey algorithm and the Guruswami-Sudan algorithm, usually operate on the symbol
level to make use of the nice algebraic properties of RS codes. Soft-decision decoders
make use of the channel reliability information. In case the code is sent over a binary
input channel, then the decoder is often a bit-level decoder. With the recent advances
in soft-decision decoding of RS codes, it was vital to benchmark the performance of
such algorithms against the optimum soft-decision maximum-likelihood decoder.

A significant amount of research has been recently devoted to finding tight bounds
on the performance of linear codes under maximum-likelihood decoding [97]. The
maximum-likelihood performance of linear codes requires the knowledge of the weight
enumerator. Unfortunately, knowing the weight enumerator of the binary images of
RS codes is very hard. Some attempts have been successful in giving the binary weight
enumerator for particular realizations of RS codes [67]. Other researchers considered
enumerating the codewords by the number of symbols of each kind in each codeword
[15]. The average binary weight enumerators of a class of generalized Reed-Solomon
codes, derived from an original RS code either by using a different basis to expand each
column in the RS generator matrix into a binary representation or by multiplying each
column in the RS generator matrix by some nonzero element in the field, were studied
by Retter [94].

One of the main motivations behind this chapter was the following question:

How can one analyze the maximum-likelihood performance of the binary images of RS codes?

In Section 2.2, we attempt to answer this question by studying the weight enumerator of the ensemble of binary images of Reed-Solomon codes. In fact we show that the ensemble weight enumerator approaches that of a random code with the same dimension. It is also well known that the minimum distance of a linear code provides a lot of insight about its performance. This motivated us to study the minimum distance of the ensemble of binary images of RS codes (Section 2.3). We show that the ensemble has an asymptotically good minimum distance. Given this result, one can search for good codes within the ensemble of binary images of Reed-Solomon codes. We then attempt to answer the above question in Section 2.4, where we analyze the performance of soft and hard-decision maximum-likelihood decoding of the binary images of the RS code. We show that the bounds developed using the techniques in this chapter are indeed tight. In Section 2.5, we conclude this chapter and highlight its main results.

2.1 Preliminaries

Given a code $\mathcal{C}$ of length n, the weight enumerator of $\mathcal{C}$ is [1]

$$E_{\mathcal{C}}(w) = |\{\boldsymbol{c} \in \mathcal{C} : \mathcal{W}(\boldsymbol{c}) = w\}|, \tag{2.1}$$

where $\mathcal{W}(\boldsymbol{c})$ is the Hamming weight of $\boldsymbol{c}$. The weight generating function (WGF) of $\mathcal{C}$ is the polynomial

$$\mathbb{E}_{\mathcal{C}}(\mathcal{X}) = \sum_{h=0}^{n} E_{\mathcal{C}}(h)\mathcal{X}^h, \tag{2.2}$$

where the coefficient of $\mathcal{X}^h$ is the number of codewords with weight h;

$$E_{\mathcal{C}}(h) = \mathrm{Coeff}\left(\mathbb{E}_{\mathcal{C}}(\mathcal{X}), \mathcal{X}^h\right). \tag{2.3}$$

(The subscript $\mathcal{C}$ may be dropped when there is no ambiguity about the code.)

For an (n, k, d) MDS code over $\mathbb{F}_q$, it is well known that the minimum distance is

---

[1] Unless otherwise noted, $|\mathcal{S}|$ is the cardinality of the set $\mathcal{S}$.

$d = n - k + 1$ [75] and that the weight distribution is given by [109, Theorem 25.7]

$$E(i) = \binom{n}{i} \sum_{j=d}^{i} \binom{i}{j} (-1)^{i-j} (q^{j-d+1} - 1) \tag{2.4}$$

$$= \binom{n}{i} (q-1) \sum_{j=0}^{i-d} (-1)^j \binom{i-1}{j} q^{i-j-d}, \tag{2.5}$$

for weights $i \geq d$.

2.2 Average Binary Image of Reed-Solomon Codes

The binary image $\mathcal{C}^b$ of an (n,k) code $\mathcal{C}$ over F_{2^m} is obtained by representing each symbol by an m-dimensional binary vector in terms of a basis of the field [75]. The weight enumerator of $\mathcal{C}^b$ will vary according to the basis used. In general, it is also hard to know the weight enumerator of the binary image of a certain Reed-Solomon code obtained by a specific basis representation (e.g., [67, 15]). For performance analysis, one could average the performance over all possible binary representations of $\mathcal{C}$. By assuming that the all such representations are equally probable, it follows that the distribution of the bits in a nonzero symbol follows a binomial distribution and the probability of having i ones in a nonzero symbol is $\frac{1}{2^m - 1} \binom{m}{i}$. The generating function of the *average* weight enumerator of the binary image of a nonzero symbol is

$$F(\mathcal{Z}) = \sum_{i=1}^{m} \frac{1}{2^m - 1} \binom{m}{i} \mathcal{Z}^i = \frac{(1 + \mathcal{Z})^m - 1}{2^m - 1}, \tag{2.6}$$

where the power of x denotes the binary weight and the all zero vector is excluded since the binary weight of a nonzero symbol is at least one. Suppose a codeword has w nonzero symbols, and the distribution of the ones and zeros in each symbol is independent from other symbols, then the possible binary weight, b, of this codeword ranges from w to mw. Since there are $E(w)$ codewords with symbol Hamming weight

w, then the *average binary* weight generating function can be derived by

$$\tilde{E}_{\mathcal{C}^b}(\mathcal{X}) \;=\; \sum_{b=0}^{nm} \tilde{E}(b)\mathcal{X}^b \tag{2.7}$$

$$=\; \mathbb{E}_{\mathcal{C}}(\mathcal{X})\big|_{\mathcal{X}:=F(\mathcal{X})} \tag{2.8}$$

$$=\; \sum_{h=0}^{n} \frac{E(h)}{(2^m-1)^h}\,\big((1+\mathcal{X})^m-1\big)^h. \tag{2.9}$$

A closed form formula for the average binary weight enumerator (BWE) is

$$\tilde{E}(b) \;=\; \mathrm{Coeff}\left(\tilde{E}_{\mathcal{C}^b}(\mathcal{X}), \mathcal{X}^b\right) \tag{2.10}$$

$$=\; \sum_{w=d}^{n} \frac{E(w)}{(2^m-1)^w} \sum_{j=0}^{w}(-1)^{w-j}\binom{w}{j}\binom{jm}{b}; \quad b \ge d. \tag{2.11}$$

These results apply to any maximum distance separable code defined over $\mathbb{F}_q$, where $q = 2^m$ and not necessarily an RS code. Widely used RS (MDS) codes have a code length $n = 2^m - 1$. In such a case the BWE derived in (2.10) agrees with the average BWE of a class of GRS codes [94]. In other words two ensembles have the same weight enumerator; the first ensemble is the ensemble of all possible binary images of a specific RS code, the second ensemble is the binary image (with a specific basis representation) of the ensemble of generalized RS codes derived from the original RS code by multiplying each column in the generator matrix by some nonzero element in the field. It is easy to see that $G_o = 1$ and that $\tilde{E}(b) = 0$ for $0 < b < d$.

By substituting for $E(w)$, for $b \ge d$, the binary weight enumerator (BWE) is given by

$$\tilde{E}(b) = (q-1) \sum_{w=d}^{n} \left(\frac{q}{q-1}\right)^w \binom{n}{w}$$

$$\sum_{v=0}^{w-d}(-1)^v \binom{w-1}{v}\left[\sum_{j=\lceil b/m \rceil}^{w}(-1)^{w-j}\binom{w}{j}\binom{jm}{b}q^{-(d+v)}\right]. \tag{2.12}$$

Although it is easy to evaluate the above formula, the term $\binom{jm}{b}$ may diverge

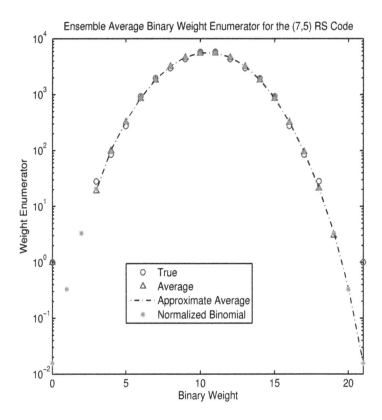

Figure 2.1: True BWE versus the averaged BWE for the $(7,5)$ RS code over $\mathbb{F}_8$.

numerically for large j. Using the Stirling approximation for $\binom{jm}{b}$ [74], $\tilde{E}(b)$ could be approximated as

$$\tilde{E}(b) \approx \sum_{w=d}^{n}(q-1)\left(\frac{q}{q-1}\right)^w \binom{n}{w}\sum_{v=0}^{w-d}(-1)^v\binom{w-1}{v}\sum_{j=\lceil b/m\rceil}^{w}\mathcal{F}(j), \qquad (2.13)$$

where

$$\mathcal{F}(j) = \begin{cases} (-1)^{w-j}\binom{w}{j}2^{\lambda(j)}, & j > b/m \\[2mm] (-1)^{w-j}\binom{w}{j}2^{-m(d+v)}, & j = b/m \end{cases}, \qquad (2.14)$$

and $\lambda(j) = m(jH(\psi_{b,j}) - d - v) - \frac{1}{2}\log_2\left(2\pi jm\psi_{b,j}(1-\psi_{b,j})\right)$ for $\psi_{b,j} = b/jm$ and $q = 2^m$. These bounds could be further simplified (and thus loosened) by observing that for $n \leq q-1$,

$$1 \leq \left(\frac{q}{q-1}\right)^w \leq \left(\frac{q}{q-1}\right)^{q-1} \leq \lim_{q\to\infty}\left(\frac{q}{q-1}\right)^{q-1} = \mathbf{e}, \qquad (2.15)$$

and substituting in (2.13).

In Figure 2.1, the averaged BWE and the true BWE for a specific basis representation found by computer search are plotted for the $(7,5)$ RS code over $\mathbb{F}_8$. The average weight enumerator of (2.12) is labeled "Average" while the approximation of (2.13) is labeled "Approximate Average." It is observed that a good approximation of the average binary weight enumerator for $h \geq d$ is the normalized binomial distribution which corresponds to a random code with the same dimension over $\mathbb{F}_q$

$$\tilde{E}(h) \approx q^{-(n-k)}\binom{mn}{h}. \qquad (2.16)$$

This observation can be somehow justified by the central limit theorem, where the binary weight of a codeword is a random variable which is the sum of n independent random variables corresponding to the binary weights of the symbols. For large n, the distribution of the binary weight is expected to converge to that of random codes. The following theorem shows that the average BWE can be upper bounded by a $\left(\frac{q}{q-1}\right)^{(n-k)}$ multiple of the above approximation.

Theorem 2.1. *The average binary weight enumerator is upper bounded by*

$$\tilde{E}(h) \leq (q-1)^{-(n-k)} \binom{mn}{h}.$$

Proof. An upper bound on the symbol weight enumerator of an (n, k, d) MDS code defined over $\mathbb{F}_q$ is [79, (12)]

$$E(w) \leq \binom{n}{w}(q-1)^{w-d+1}; \quad w \geq d. \tag{2.17}$$

Substituting in (2.10) it follows that for $b \geq d$

$$\tilde{E}(b) \leq (q-1)^{k-n} \sum_{w=d}^{n} \binom{n}{w} \left[\sum_{j=\lceil b/m \rceil}^{w} (-1)^{w-j} \binom{w}{j}\binom{jm}{b} \right]. \tag{2.18}$$

By doing a change of variables $\alpha = mj$ and changing the order of summations

$$
\begin{aligned}
\tilde{E}(b) \quad &\leq \quad (q-1)^{k-n} \sum_{w=d}^{n} \sum_{\alpha=b}^{mw} (-1)^{w-j} \binom{n}{w}\binom{w}{\alpha/m}\binom{\alpha}{b} \\
&= \quad (q-1)^{k-n} \sum_{\alpha=b}^{nm} (-1)^{-\frac{\alpha}{m}} \binom{\alpha}{b} \sum_{w=\max(\frac{\alpha}{m}, d)}^{n} (-1)^{w} \binom{n}{w}\binom{w}{\alpha/m} \\
&\leq \quad (q-1)^{k-n} \sum_{\alpha=b}^{nm} (-1)^{-\frac{\alpha}{m}} \binom{\alpha}{b} \sum_{w=\frac{\alpha}{m}}^{n} (-1)^{w} \binom{n}{w}\binom{w}{\alpha/m}.
\end{aligned}
$$

From the identity $\binom{n}{m}\binom{m}{p} = \binom{n}{p}\binom{n-p}{m-p}$ it follows that $\sum_{k=m}^{n}(-1)^k \binom{n}{k}\binom{k}{m} = (-1)^m \delta_{nm}$ where $\delta_{n,m}$ is the Kronecker delta function. It follows that

$$
\begin{aligned}
\tilde{E}(b) \quad &\leq \quad (q-1)^{k-n} \sum_{\alpha=b}^{nm} \binom{\alpha}{b} \delta_{\frac{\alpha}{m}, n} \\
&= \quad (q-1)^{k-n} \binom{mn}{b},
\end{aligned}
$$

which completes the proof.   □

In Figure 2.2, we plot the ensemble average weight enumerator of (2.10) and com-

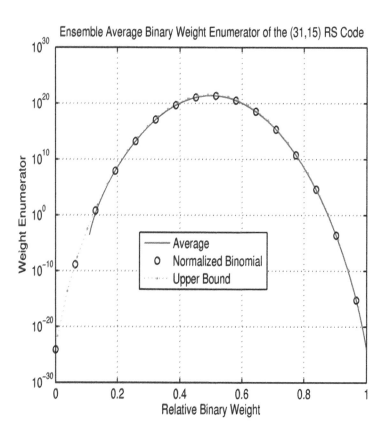

Figure 2.2: The ensemble weight enumerator of the $(31, 15)$ RS code over $\mathbb{F}_{32}$. The ensemble average weight enumerator of (2.10) is compared with the weight enumerator of the random code (2.16) and the upper bound of Theorem 2.1. They are labeled "Average," "Normalized Binomial" and "Upper Bound" respectively.

pare it with the weight enumerator of a random code with the same dimension (2.16). We also compare it with the simple upper bound of Theorem 2.1. It is observed that the upper bound of Theorem 2.1 is fairly tight and that a good approximation for the ensemble weight enumerator is that of random codes. In fact, as length of the code (and the size of the finite field) tend to infinity

$$\tilde{E}(h) \;\leq\; \left(\frac{q}{q-1}\right)^{(n-k)} q^{-(n-k)} \binom{mn}{h} \tag{2.19}$$

$$\leq\; \mathbf{e} 2^{-m(n-k)} \binom{mn}{h} \tag{2.20}$$

$$\leq\; \frac{\mathbf{e}}{\sqrt{2\pi mn\lambda(1-\lambda)}} \, 2^{mn(H_2(\lambda)-1+R)}, \tag{2.21}$$

where $b = \lambda mn$, $R = k/n$ is the code rate and $H_2(\lambda)$ is the binary entropy function. The last inequality follows from the Stirling's inequality [74, p. 309]. Let the asymptotic weight enumerator exponent of a code $\mathcal{C}$, of length N and weight enumerator $E_{\mathcal{C}}$, be defined as

$$\Xi(\lambda) \;\overset{\triangle}{=}\; \lim_{N\to\infty} \frac{\log_2\left(E_{\mathcal{C}}(\lambda N)\right)}{N}. \tag{2.22}$$

It follows that the asymptotic weight enumerator exponent of the ensemble of binary images of Reed-Solomon codes is

$$\begin{aligned}
\tilde{\Xi}(\lambda) \;&=\; \lim_{\substack{n\to\infty \\ m\to\infty}} \frac{\log_2\left(\tilde{E}(\lambda mn)\right)}{mn} \\
&\leq\; \lim_{\substack{n\to\infty \\ m\to\infty}} \frac{\log_2(\mathbf{e}) - \frac{1}{2}\log_2(mn) - \frac{1}{2}\log_2(2\pi\lambda(1-\lambda))}{mn} + H_2(\lambda) - 1 + R \\
&=\; H_2(\lambda) - (1-R). \tag{2.23}
\end{aligned}$$

In other words, as the code length and the finite field size tend to infinity, the weight enumerator of the ensemble of binary images of an RS code approaches that of a random code.

The error-correcting capability of a code relies a lot on the minimum distance of the code, which will be analyzed in the next section.

2.3   The Binary Minimum Distance of the Ensemble of Binary Images of Reed-Solomon Codes

The error-correcting capability of a code relies a lot on the minimum distance of the code. We will now consider the minimum distance of the ensemble of binary images of a certain Reed-Solomon code. The average minimum distance of the binary image of the RS code could be defined to be the smallest weight b whose average BWE $\tilde{E}(b)$ is greater than or equal to one (note that $\tilde{E}(b)$ is a real number). Let d_b be the average BMD, then

$$d_b \triangleq \inf_{b \geq d}\{b : \tilde{E}(b) \geq 1\}. \tag{2.24}$$

The number d_b could be found exactly by numerical search. However, it will also be useful to find a lower bound on d_b. It is straightforward to note that the binary minimum distance (BMD) is at least as large as the symbol minimum distance d;

$$d_b \geq n - k + 1. \tag{2.25}$$

In the following theorems, we will give some lower bounds on the average binary minimum distance of the ensemble of binary images.

Theorem 2.2. *The minimum distance of the ensemble of binary images of an (n, k, d) RS code over $\mathbb{F}_{2^m}$ is lower bounded by*

$$d_b \geq \inf_{b \geq d}\left\{b : \binom{mn}{b} \geq (2^m - 1)^{n-k}\right\}.$$

Proof. From the upper bound on $\tilde{E}_b$ of Theorem 2.1, and the definition of d_b, the theorem follows. □

Theorem 2.3. *A lower bound on d_b is*

$$d_b \geq \inf_{b \geq d}\left\{b : \sum_{w=d}^{n} \binom{n}{w}\binom{wm}{b} \geq (2^m - 1)^{n-k}\right\}.$$

Proof. By taking only the term corresponding to $j = w$ in the alternating sign summation in (2.18), it follows that

$$\tilde{E}(b) \leq (q-1)^{k-n} \sum_{w=d}^{n} \binom{n}{w}\binom{wm}{b}.$$

The theorem follows from the definition of d_b.   □

Since the upper bound on the weight enumerator of (2.26) is not tighter than the bound of Theorem 2.1, it is expected that the lower bound on the minimum distance of Theorem 2.3 will not be tighter than that of Theorem 2.2.

Since the binary minimum distance of the ensemble is at least as large as the symbol minimum distance (c.f., (2.25)), it is interesting to determine when the binary minimum distance is equal to the symbol minimum distance which is linear in the rate R of the code.

Lemma 2.4. *The average binary minimum distance of an MDS code over $\mathbb{F}_{2^m}$ is equal to its symbol minimum distance for all rates greater than or equal to $R_o = 1 - \frac{d_o - 1}{n}$ where d_o is the largest integer d' such that*

$$\frac{1}{d'} \log_2\left((2^m - 1)\binom{n}{d'}\right) \geq \log_2(2^m - 1) - \log_2(m). \tag{2.26}$$

Proof. The number of codewords in an MDS code with symbol weight $d = n - k + 1$ is $E(d) = (q-1)\binom{n}{d}$. The binary image could be of binary weight d only if the codeword is of symbol weight d and the binary representation of each nonzero symbol has only one nonzero bit. This happens with probability $\left(\frac{m}{2^m - 1}\right)^d$, where $m = \log_2(q)$. So the average number of codewords with binary weight d is

$$\tilde{E}(d) = E(d)\left(\frac{m}{2^m - 1}\right)^d = (q-1)\binom{n}{d}\left(\frac{\log_2(q)}{q-1}\right)^d. \tag{2.27}$$

From the definition of the average binary minimum distance, the lemma follows.   □

Asymptotically, it could be shown that R_o is the smallest rate such that

$$\frac{H_2(1 - R_o)}{(1 - R_o)} \geq \log_2(n) - \log_2(\log_2(n)), \tag{2.28}$$

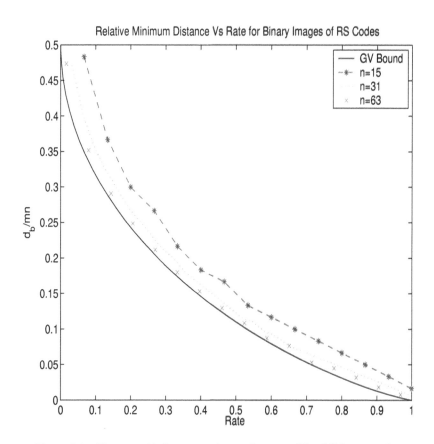

Figure 2.3: The ensemble binary minimum distance of Reed-Solomon codes. The Relative binary minimum distance for the ensemble of binary images of Reed-Solomon codes is plotted against the rate for lengths 15, 31 and 63 over finite fields of sizes 16, 32 and 64 respectively and compared with the Gilbert-Varshamov bound.

where $n \approx q$ and

$$H_2(x) = -x \log_2(x) - (1 - x) \log_2(1 - x) \tag{2.29}$$

is the binary entropy function. This implies that the rate R_o, at which the symbol minimum distance is equal to the ensemble binary minimum distance, tends to one as the length of the code tends to infinity.

The Gilbert-Varshamov (GV) bound is defined by [74],

$$\lim_{n \to \infty} \{R(\delta) - (1 - H_2(\delta))\} \geq 0 \text{ for } 0 < \delta < \frac{1}{2}, \tag{2.30}$$

where $\delta = d_b/(mn)$ is the ratio of the binary minimum distance to the total length of the code and $R(\delta)$ is rate of the code with a relative minimum distance δ. Retter showed that for sufficiently large code lengths, most of the codes in the binary image of the ensemble of generalized RS codes lie close to the GV bound by showing that the number of codewords with weights lying below the GV bound in all generalized RS codes of the same length and rate are less than half the number of such generalized RS codes [94]. Next, we show a related result for the ensemble of binary images of an RS code, with a binary weight enumerator $\tilde{E}(b)$.

We will now determine a bound on the asymptotic relative binary minimum distance (as the length tends to infinity) of the ensemble of binary images, δ_∞

$$\delta_\infty \stackrel{\Delta}{=} \inf_\lambda \{\tilde{\Xi}(\lambda) \geq 0\}. \tag{2.31}$$

From the asymptotic analysis of (2.23), we showed that

$$\tilde{\Xi}(\lambda) \leq H_2(\lambda) - (1 - R). \tag{2.32}$$

It thus follows that

$$\delta_\infty \geq \inf_\lambda \{H_2(\lambda) \geq (1 - R)\}. \tag{2.33}$$

One can then deduce that

$$H_2(\delta_\infty) - (1 - R(\delta_\infty)) \geq 0. \tag{2.34}$$

In other words, we have proved the following theorem,

Theorem 2.5. *The ensemble of binary images of an Reed-Solomon code asymptotically satisfies the Gilbert-Varshamov bound.*

This is not very surprising since we have shown that the ensemble average behaves like a binary random code. Note that this is for the average binary image of the RS code and not for a specific valid binary image. Since this theorem is for the ensemble average, it might imply that some codes in the ensemble may have a minimum distance asymptotically satisfying the GV bound. However, we do not know of a specific code in the ensemble that satisfies the bound.

In Figure 2.3, we show the relative average binary minimum distance for binary images of Reed-Solomon codes, calculated numerically by (2.24), for different code lengths. It is observed that as the length and the size of the finite field increases, the relative minimum distance decreases. From Theorem 2.5, the relative binary minimum distance of the ensemble should approach the GV bound as the length tends to infinity. In Figure 2.4 and Figure 2.5, we study the relative average binary minimum distance for code lengths $n = 15$ and $n = 31$ respectively. We compare it with the Gilbert-Varshamov bound and the lower bounds of Theorem 2.2 and the linear bound of (2.25). We observe that the lower bound of Theorem 2.2 is pretty tight and it provides a simple way to evaluate the minimum distance of the ensemble. Moreover it is always lower bounded by the GV bound. By comparing with the linear lower bound of (2.25), it is noticed that for $n = 15$ and $k \geq 8$, the average BMD is equal to the symbol minimum distance, d, as expected from Lemma 2.4. As the rate decreases, this linear lower bound becomes very loose and the average binary minimum distance exceeds the symbol minimum distance.

2.4 Performance of the Maximum-Likelihood Decoders

Let c be the binary image of a codeword in the (n, k, d) RS code $\mathcal{C}$. The binary phase shift keying (BPSK) modulated image of c is $x = \mathcal{M}(c) = 1 - 2c$. This will

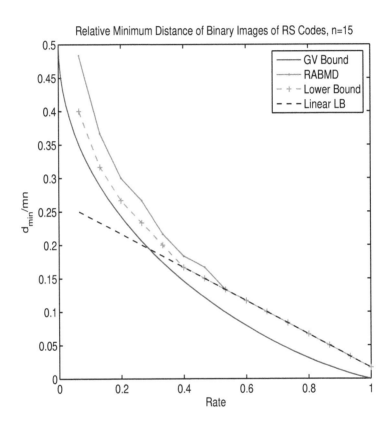

Figure 2.4: The ensemble binary minimum distance of RS codes of length 15 over $\mathbb{F}_{16}$. The relative binary minimum distance is plotted versus the code rate. The numerical minimum distance (2.24) is labeled "RABMD" and compared with the lower bounds of Theorem 2.2 and (2.25) which are labeled "Lower Bound" and "Linear LB" respectively. The Gilbert-Varshamov bound is plotted and labeled "GV Bound."

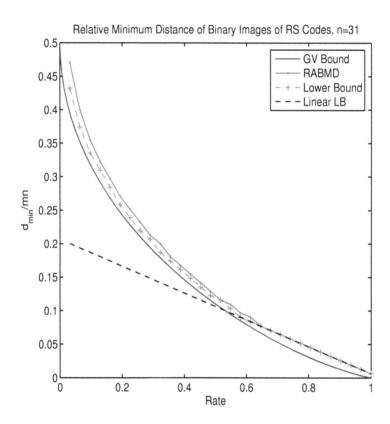

Figure 2.5: The ensemble binary minimum distance of RS codes of length 31 over $\mathbb{F}_{32}$. The relative binary minimum distance for the ensemble of binary images of the RS codes are plotted versus the code rate. The numerical minimum distance (2.24) is labeled "RABMD" and compared with the lower bounds of Theorem 2.2 and (2.25) which are labeled "Lower Bound" and "Linear LB" respectively. The Gilbert-Varshamov bound is plotted and labeled "GV Bound."

be transmitted over a standard binary input additive white Gaussian noise (AWGN) channel. The received vector is $y = x + z$, where z is an AWGN vector. Since the considered codes are linear, it is safe to assume that the all zero codeword (in fact its binary image) is transmitted. Hard-decision decoding is done to the received bits to obtain the vector $\bar{y}$ where $\bar{y}_i = (1 - \text{sign}(y_i))/2$ and the HD-ML decoder's output is the codeword $\hat{c}$ such that

$$\hat{c} = \arg \min_{v \in \mathcal{C}^b} d(\bar{y}, v), \tag{2.35}$$

where $d(u, v)$ is the (binary) Hamming distance between u and v. This is equivalent to transmitting the codeword c through a binary symmetric channel (BSC) with crossover probability $p = Q(\sqrt{2R\gamma})$ where γ is the bit signal-to-noise ratio and R is the code rate.

As discussed before, bounds on the error probability of linear codes require the knowledge of the weight enumerator. For a specific binary image, it is very hard to know the weight enumerator. It is also hard to agree on the use of a specific binary image or to speculate which binary image has been used. So the question we really need to answer is the expected performance if any binary image of a specific RS code is used. Our approach is to consider the binary code of a weight enumerator equal to the ensemble average weight enumerator.

The performance of the hard-decision maximum-likelihood (HD-ML) decoder can be upper bounded with the well-known union bound by resorting to the average weight enumerator of the ensemble

$$P(\mathcal{E}_{HML}) \leq \sum_{b=d_b}^{mn} \tilde{E}(b) \sum_{w=\lceil \frac{b}{2} \rceil}^{b} \binom{b}{w} p^w (1 - p)^{b-w}, \tag{2.36}$$

where $P(\mathcal{E}_{HML})$ denotes the codeword error probability of the HD-ML decoder. Alternatively, one could use the ensemble average weight enumerator with tighter bounds. The best well-known upper bound on the performance of a HD-ML decoding of linear codes on binary symmetric channels is the Poltyrev bound [87] (c.f., (8.32)).

The soft-decision maximum-likelihood decoder solves the following optimization problem,

$$\hat{c} = \arg \min_{\boldsymbol{v} \in \mathcal{C}_b} \|\boldsymbol{y} - \mathcal{M}(\boldsymbol{v})\|^2 \tag{2.37}$$

where $\|\boldsymbol{x}\|$ is the Euclidean norm of $\boldsymbol{x}$. Assuming that the all-zero codeword is BPSK modulated and transmitted over a memoryless AWGN channel, the probability that a certain codeword of binary weight b is chosen at the decoder instead of the transmitted all-zero codeword is [89, (8.1-49)] $P_b = Q\left(\sqrt{2\gamma Rb}\right)$, where γ is the signal-to-noise ratio (SNR) per bit and $R = k/n$.

Then a heuristic union lower bound on the codeword error probability of the soft-decision maximum-likelihood decoder (specifically true at high SNRs) is the probability that a codeword of minimum weight d_b is erroneously decoded,

$$P(\mathcal{E}_{SML}) \gtrsim \tilde{E}(d_b)Q\left(\sqrt{2\gamma Rd_b}\right). \tag{2.38}$$

A union upper bound on the codeword error probability is the sum of all possible errors,

$$P(\mathcal{E}_{SML}) \leq \sum_{b \geq d_b} \tilde{E}(b)Q\left(\sqrt{2\gamma Rb}\right). \tag{2.39}$$

The union bound is loose at low SNRs. Poltyrev described a tangential sphere bound (TSB) on the error probability of binary block codes BPSK modulated in AWGN channels [87]. This is a very tight upper bound on the ML error probability. For a brief description of the Tangential Sphere Bound we refer the reader to Section 8.1.3. We use the TSB in conjunction with the average binary weight enumerator to find a tight upper bound on the error probability of ML decoding of RS codes. Divsalar also introduced in [23] a simple tight bound (that involves no integrations) on the error probability of binary block codes, as well as a comparison of other existing bounds. Other bounds such as the variations on the Gallager bounds are also tight for AWGN and fading channels [99].

The Berlekamp-Massey (BM) decoder is a symbol-based hard-decision decoder which can correct a number of symbol errors up to half-the-minimum distance of the code, $\tau_{BM} = \lfloor \frac{n-k}{2} \rfloor$. The error plus failure probability of the BM decoder has been well

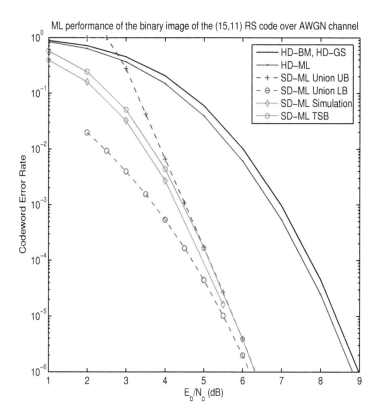

Figure 2.6:   Performance of a binary image of $(15, 11)$ RS code over $\mathbb{F}_{16}$ when transmitted over a binary input AWGN channel.

The analytic performance of the symbol-level hard-decision Berlekamp-Massey and Guruswami-Sudan decoders are shown and are labeled by "HD-BM" and "HD-GS" respectively.   These are in turn compared to the bit-level HD ML decoder labeled "HD-ML." The union upper bound (2.39), lower bound (2.38) and the tangential sphere bound on the soft-decision ML error probability are labeled "SD-ML Union UB," "SD-ML Union LB" and "SD-ML TSB" respectively. The simulated performance of an SD ML decoder is labeled "SD-ML Simulation."

studied [79, 115] and can be simply given by

$$P(\mathcal{E}_{BM}) = 1 - \sum_{j=0}^{\tau_{BM}} \binom{n}{j} (1-s)^j s^{n-j},$$

where s is the probability that a symbol is correctly received $s = \left(1 - Q\left(\sqrt{2\gamma R}\right)\right)^m$. The Guruswami-Sudan decoder is also a symbol-based HD decoder but can correct more than half-the-minimum distance of the code $\tau_{GS} = \lceil n - \sqrt{nk} - 1 \rceil$. The performance of a hard-decision "sphere" decoder that corrects any number of $\tau \geq \tau_{BM}$ symbol errors as well that of the corresponding maximum-likelihood decoder over q-ary symmetric channels have been recently analyzed [37, 36].

We evaluate the average performance of RS codes when its binary image is BPSK modulated and transmitted over an AWGN channel. In Figure 2.6, we consider a specific binary image of the $(15,11)$ RS code over $\mathbb{F}_{16}$. Soft-decision maximum-likelihood decoding was simulated using the BCJR algorithm [7] on the trellis associated with the binary image of the RS code [66]. By comparing this with the average TSB, we observe that our technique for bounding the performance of the soft-decision ML decoder provides tight upper bounds on the actual performance of a specific binary image. It is clear that at low SNRs the (averaged) TSB give a close approximation of the ML error probability. By comparing this bound with the union upper and lower bounds of (2.39) and (2.38), we observe that the TSB coincides with the union bounds at high SNRs. As from (2.38), the union lower bound is characterized by the minimum distance term. Indeed, the SNR at which the performance of the maximum-likelihood decoder is dominated by the minimum distance term was recently studied by Fossorier and was termed the *critical point* for ML decoding [44]. The decoding radius of the GS decoder is the same as that of the BM decoder for the $(15,11)$ code, which is of relatively high rate. However, their performance is very close to that of the HD-ML decoder.

In Figure 2.7, we consider the performance of the binary image of the $(31,15)$ RS code over $\mathbb{F}_{16}$ when BPSK modulated and transmitted over an AWGN channel. We compare the performance of a bit-level HD-ML decoder with that of a symbol-level HD-ML decoder by deploying the bounds of [87] and [36] respectively. The symbol-

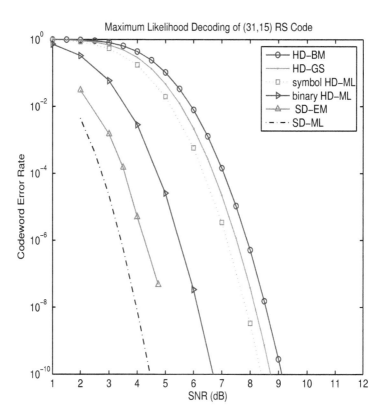

Figure 2.7: Performance of the binary image of the $(31, 15)$ RS code over $\mathbb{F}_{32}$ transmitted over a binary input AWGN channel.

The symbol-level HD-BM and the HD-GS algorithms are compared. Bit-level and symbol-level hard-decision decoders are labeled "binary HD-ML" and "symbol HD-ML" respectively. The TSB on the bit-level SD-ML error probability is labeled "SD-ML" and is compared with the bit-level soft-decision algorithm of [33] labeled "SD-EM."

level decoder operates by first grouping m bits to symbols in $\mathbb{F}_{2^m}$ after hard-decision decoding. It seems that for this half-rate code, the performance of a bit-level HD decoder is better than the corresponding symbol-level decoder (about 1.5 dB coding gain). We also compare the performance with that of the symbol-level HD-BM and the HD-GS algorithms. For the $(31, 15)$ code, bit-level HD-ML decoding has more than 2 dB gain over the BM decoder, whereas SD-ML decoding offers another 2 dB gain over bit-level HD-ML decoding. The SD-ML decoder has about 4 dB gain over the BM decoder and 2 dB gain over the HD-ML decoder. Bounds on the performance of the maximum-likelihood decoder provides a benchmark to compare the performance of other suboptimum algorithms. To emphasize this, the performance of a bit-level soft-decision decoder, developed in Chapter 5, acting on a specific binary image is also plotted. Only by comparing it to the SD-ML bound can one conclude that this soft-decision algorithm operates within 1 dB of the optimum soft-decision algorithm.

2.5 Conclusion

An averaged binary weight enumerator for RS codes is derived and shown to closely estimate an exact binary weight enumerator for a specific basis representation. Moreover, it has been shown that as the code length and the field size tend to infinity, the weight enumerator of the ensemble of binary images of Reed-Solomon codes approach that of a random code with the same dimensions. Bounds on the average binary minimum distance were derived. It was thus shown that on average, the ensemble of binary images of RS codes asymptotically satisfy the GV bound. The question remains open, if there exists a specific code in the ensemble that asymptotically satisfies the GV bound. Aided with the ensemble weight enumerator, one can derive tight bounds on the performance of bit-level maximum-likelihood decoders. By comparing with simulations, it has been shown, that at least for the $(15, 11)$ RS code, the tangential sphere bound when combined with the ensemble weight enumerator is tight. When proposing new algorithms for decoding RS codes, it is not only important to compare their performance with other algorithms in the literature, but it is also more important to compare their performance with that of other maximum-likelihood decoders using the results in this chapter.

Chapter 3

The Multiuser Error Probability of Reed-Solomon Codes

> *Not everything that can be counted counts, and not everything that counts can be counted.*
>
> —Albert Einstein

Consider a network scenario, where users in a certain cluster can communicate in an error free manner. These users would like to communicate with another set of users in another cluster over a noisy channel. If the users in the first cluster are of limited power they will not be able to reliably transmit their information to the users in the other cluster. One solution is for the users to transmit their information to a local base station, which will then group their data symbols, encode them with a channel code and transmit the codeword to the other set of users (see Figure 3.1). In other words, each codeword will be partitioned among more than one user or application. After decoding at the receiving base station, the information will be routed to the desired users. One other advantage of sharing a codeword among different users is the expected improvement in the code performance as its length increases [25]. Moreover, the recent results on the capacity of wireless networks suggest that networks with a smaller number of users and clustered networks are more likely to find acceptance [47]. Using the results in this chapter, we will be able to analyze the performance of different users in such a scenario when the code is a maximum distance separable (MDS) code.

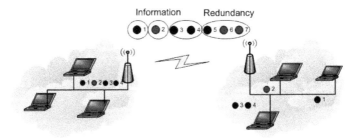

Figure 3.1: A multiuser scenario where a code is shared among many users. Users within the same cluster transmit their information to a local base station, which, in turn, groups their symbols into one data word and transmits it, after channel encoding, over a noisy channel to the users in another cluster.

Reed-Solomon codes are the most widely used MDS codes. The results here can also be useful in the analysis of MDS codes in distributed storage systems, where MDS array were proposed [16].

In Section 3.1, we introduce a generalized weight enumerator, which we call the partition weight enumerator (PWE). Given a partition of the coordinates of a code, the PWE enumerates the codewords with a certain weight profile in the partitions. Our main result is a simple closed-form expression for the PWE of an arbitrary MDS, e.g., Reed-Solomon, code (Section 3.2, Theorem 3.6). This generalizes the results of Kasami *et al.* [69] on the split weight enumerator of RS codes. The PWE is a very useful tool in proving some of the nice algebraic properties of MDS codes. We then proceed in Section 3.3 to derive a strong symmetry property for MDS codes (Theorem 3.8) which allows us to obtain improved bounds on the symbol error probability for RS codes. We show that an approximation widely used to estimate the symbol error probability of linear codes is exact for MDS codes. We take this opportunity to discuss other codes which also have this property.

As we have mentioned in Chapter 2, the ensemble average weight enumerators of the binary images of RS codes have been rendered useful in analyzing their performance. We also study the case when the binary images of an Reed-Solomon code is partitioned among different users or applications. In Section 3.4, we show that the ensemble also has a similar symmetry property which becomes useful when analyzing its bit error probability.

We study, in Section 3.5, the codeword, symbol and bit error probabilities of various Reed-Solomon code decoders in a generalized setting. In Section 3.6, we prove that if systematic MDS (e.g., RS) codes are used in a multiuser setting, the unconditional symbol or bit error probabilities of all the users will be the same regardless of the size of the partitions assigned to them. We also considered various network scenarios where the Reed-Solomon code is the channel code of choice. We also proceed to show how one can analyze the error probability of a certain user given some conditions on the performance of other users. In Section 3.7, we conclude the chapter and give some insights about the results in this chapter.

3.1 Weight Enumerators

We begin by generalizing the notion of Hamming weight. Let $\mathbb{F}_q^n$ denote the vectors of length n over the finite field of q elements $\mathbb{F}_q$. A linear code $\mathcal{C}$ of length n defined over $\mathbb{F}_q$ is a linear subspace of $\mathbb{F}_q^n$. Let $N = \{1, 2, \ldots, n\}$ be the coordinate set of $\mathcal{C}$. Suppose N is partitioned into p disjoint subsets $N_1, \ldots, N_p$, with $|N_i| = n_i$, for $i = 1, \ldots, p$. [1] We stress that $\sum_{i=1}^{p} n_i = n$. The elements of the set $N_i \subset N$ are given by $N_i = \{N_i(1), N_i(2), \ldots, N_i(n_i)\}$. Let $\boldsymbol{v} = (\boldsymbol{v}_1, \boldsymbol{v}_2, \ldots, \boldsymbol{v}_n)$ be a vector in $\mathbb{F}_q^n$, then the ith partition of $\boldsymbol{v}$ is the vector $\boldsymbol{v}[N_i] = \left(\boldsymbol{v}_{N_i(1)}, \boldsymbol{v}_{N_i(2)}, \ldots, \boldsymbol{v}_{N_i(n_i)}\right)$.

Note that the number of ways a set of n coordinates could be partitioned into m_1 partitions of size of p_1, m_2 partitions of size p_2 and m_r of size p_r, i.e., the total number of partitions is $\sum_{i=1}^{r} m_r$ and $n = \sum_{i=1}^{r} m_r p_r$), is

$$\frac{n!}{\prod_{i=1}^{r}(p_i!)^{m_i} m_i!},\tag{3.1}$$

where $x!$ is the factorial of x and the multinomial coefficient is normalized by the factor $\prod_{i=1}^{r} m_i!$ as we do not distinguish between partitions of the same size.

Denoting an $(n_1, \ldots, n_p)$ partition by $\mathcal{T}$, the $\mathcal{T}$-weight profile of a vector $\boldsymbol{v} \in \mathbb{F}_q^n$ is

---

[1]Throughout this chapter, the cardinality of a set T will be denoted by $|T|$.

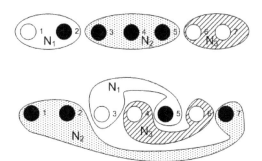

Figure 3.2: Partitioning of a code defined over $\mathbb{F}_q^7$.
The figure shows two different vectors in $\mathbb{F}_q^7$ and two different $\mathcal{T} : (2,3,2)$ partitions are applied. The weight profile of the vectors is $\mathcal{W}_\mathcal{T}(\boldsymbol{v}) = (1,3,0)$ where the zero and nonzero symbols are represented by white and black circles respectively.

defined as $\mathcal{W}_\mathcal{T}(\boldsymbol{v}) = (w_1,\dots,w_p)$, where w_i is the Hamming weight of $\boldsymbol{v}$ restricted to N_i, i.e., the weight of the vector $\boldsymbol{v}(N_i)$. (For an example see Figure 3.2.) The weight enumerator of a code $\mathcal{C}$ is as defined in (2.2).

Now we generalize the notion of code weight enumerator. For an $(n_1, n_2, \dots, n_p)$ partition $\mathcal{T}$ of the n coordinates of $\mathcal{C}$, the $\mathcal{T}$-weight enumerator of $\mathcal{C}$ enumerates the codewords with a weight profile $(w_1, \dots, w_p)$,

$$A_\mathcal{C}^\mathcal{T}(w_1,\dots,w_p) = |\{\boldsymbol{c} \in \mathcal{C} : \mathcal{W}_\mathcal{T}(\boldsymbol{c}) = (w_1,\dots,w_p)\}|.$$

The *partition weight generating function* (PWGF) is given by the multivariate polynomial

$$\mathbb{P}^\mathcal{T}(\mathcal{X}_1,\dots,\mathcal{X}_p) = \sum_{w_1=0}^{n_1} \dots \sum_{w_p=0}^{n_p} A^\mathcal{T}(w_1,\dots,w_p)\mathcal{X}_1^{w_1}\dots\mathcal{X}_p^{w_p}. \tag{3.2}$$

For the special case of two partitions, $(p = 2)$, $A^\mathcal{T}(w_1, w_2)$ is termed the *split weight enumerator* in the literature [74]. The *input-redundancy weight enumerator* (IRWE) $R(w_1, w_2)$ is the number of codewords with input weight (weight of the information vector) w_1 and redundancy weight w_2. For a systematic code, if $\mathcal{T}$ is an

$(k, n - k)$ partition such that the first partition constitutes of the coordinates of the information symbols, then $R(w_1, w_2) = A^T(w_1, w_2)$. The *input-output weight enumerator* (IOWE) $O(w, h)$ enumerates the codewords of total Hamming weight h and input weight w. Assuming that the first partition constitutes of the information symbols, then $O(w, h) = R(w, h - w)$. For an $(k, n - k)$ partition T, it is straightforward that

$$E(h) = \sum_{w=0}^{k} A^T(w, h - w) = \sum_{w=0}^{k} O(w, h). \tag{3.3}$$

It is useful to know the IOWE and IRWE of a code when studying its bit error probability (e.g., [8]). The *input-output weight generating function*, $\mathbb{O}(\mathcal{X}, \mathcal{Y})$, and the *input-redundancy weight generating function*, $\mathbb{R}(\mathcal{X}, \mathcal{Y})$, of an (n, k) code are defined to be respectively,

$$\mathbb{O}(\mathcal{X}, \mathcal{Y}) = \sum_{w=0}^{k} \sum_{h=0}^{n} O(w, h) \mathcal{X}^w \mathcal{Y}^h, \tag{3.4}$$

$$\mathbb{R}(\mathcal{X}, \mathcal{Y}) = \sum_{w_1=0}^{k} \sum_{w_2=0}^{n-k} R(w_1, w_2) \mathcal{X}^{w_1} \mathcal{Y}^{w_2}. \tag{3.5}$$

Since every nonzero symbol in the redundancy part of the code contributes to both its output and redundancy weights, $\mathbb{R}(\mathcal{X}, \mathcal{Y})$ and $\mathbb{O}(\mathcal{X}, \mathcal{Y})$ are related by the following transformations,

$$\mathbb{R}(\mathcal{X}, \mathcal{Y}) = \mathbb{O}\left(\frac{\mathcal{X}}{\mathcal{Y}}, \mathcal{Y}\right), \quad \mathbb{O}(\mathcal{X}, \mathcal{Y}) = \mathbb{R}(\mathcal{X}\mathcal{Y}, \mathcal{Y}), \quad \mathbb{E}(\mathcal{X}) = \mathbb{R}(\mathcal{X}, \mathcal{X}). \tag{3.6}$$

For a systematic code, let the jth partition consist of information symbols, then the jth IOWE enumerates the codewords with a Hamming weight w in the jth partition and a total weight h,

$$O^j(w, h) = |\{\mathbf{c} \in \mathcal{C} : (\mathcal{W}(\mathbf{c}[N_j]) = w) \wedge (\mathcal{W}(\mathbf{c}) = h)\}|, \tag{3.7}$$

and is derived from the PWGF by

$$\mathbb{O}^j(\mathcal{X},\mathcal{Y}) = \mathbb{P}^{\mathcal{T}}(\mathcal{Y},\mathcal{Y},..,\mathcal{X}\mathcal{Y},..,\mathcal{Y}) = \sum_{w=0}^{n_j}\sum_{h=0}^{n} O^j(w,h)\mathcal{X}^w\mathcal{Y}^h, \qquad (3.8)$$

where the invariants $\mathcal{X}_i$s in $\mathbb{P}_{\mathbb{C}}^{\mathcal{T}}(\mathcal{X}_1,\mathcal{X}_2,\ldots,\mathcal{X}_p)$ are substituted by

$$\begin{cases} \mathcal{X}_i := \mathcal{Y}, & \forall\ i \neq j \\ \mathcal{X}_i := \mathcal{X}\mathcal{Y}, & i = j \end{cases}. \qquad (3.9)$$

3.2 Partition Weight Enumerator of Maximum-Distance-Separable Codes

For an (n,k,d) MDS code over $\mathbb{F}_q$, it is well known that the minimum distance is $d = n - k + 1$ [75] and that the weight distribution is given by [109, Theorem 25.7]

$$E(i) = \binom{n}{i}\sum_{j=d}^{i}\binom{i}{j}(-1)^{i-j}(q^{j-d+1}-1) \qquad (3.10)$$

$$= \binom{n}{i}(q-1)\sum_{j=0}^{i-d}(-1)^j\binom{i-1}{j}q^{i-j-d}, \qquad (3.11)$$

for weights $i \geq d$. In the next theorem, we show that for an arbitrary partition of the coordinates of an MDS code, and for any number of partitions, the partition weight enumerator of MDS codes admits a closed form formula.

Theorem 3.1. *For an (n,k,d) MDS code $\mathcal{C}$ defined over $\mathbb{F}_q$, let $\mathcal{T}$ define a p-partition of the coordinates of $\mathcal{C}$ into p mutually exclusive subsets, N_1, N_2, ..., N_p, such that $N_1 \cup N_2... \cup N_p = N$ where $N = \{1,2,\ldots,n\}$ and $|N_i| = n_i$. The p-partition weight*

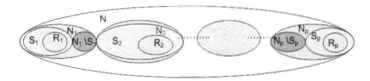

Figure 3.3: Theorem 3.1.
The code is always zero on the coordinates in the sets $N_i \setminus S_i$ for $i = 1, 2, \ldots, p$.

enumerator is given by

$$\binom{n_1}{w_1} \cdots \binom{n_p}{w_p} \sum_{j_1=0}^{w_1} \binom{w_1}{j_1} (-1)^{w_1-j_1} \sum_{j_2=0}^{w_2} \binom{w_2}{j_2} (-1)^{w_2-j_2}$$

$$\cdots \sum_{j_p=d-\sum_{z=1}^{p-1} j_z}^{w_p} \binom{w_p}{j_p} (-1)^{w_p-j_p} (q^{\sum_{z=1}^{p} j_z-d+1} - 1).$$

Proof. For $i = 1, 2, \ldots, p$, let R_i be a subset of N_i. Define $S(\boldsymbol{c})$ to be the support set of the codeword $\boldsymbol{c}$, i.e., the set of indices of the nonzero elements. Define

$$f(R_1, R_2, \ldots, R_p) \triangleq |\boldsymbol{c} \in \mathcal{C} : \{S(\boldsymbol{c}) \cap N_i\} = R_i \; \forall i| = |\boldsymbol{c} \in \mathcal{C} : \{S(\boldsymbol{c}) = \bigcup_{i=1}^{p} R_i\}| \quad (3.12)$$

to be the number of codewords which are exactly nonzero on the sets R_i. From the definition of the p-partition weight enumerator, it follows that

$$A^T(w_1, w_2, \ldots, w_p) = \sum_{\substack{R_1 \subseteq N_1 \\ |R_1|=w_1}} \sum_{\substack{R_2 \subseteq N_2 \\ |R_2|=w_2}} \cdots \sum_{\substack{R_p \subseteq N_p \\ |R_p|=w_p}} f(R_1, R_2, \ldots, R_p). \quad (3.13)$$

Define the mutually exclusive subsets, $S_i \subseteq N_i$, $i = 1, 2, \ldots, p$ and let

$$g(S_1, S_2, \ldots, S_p) = \sum_{R_1 \subseteq S_1} \sum_{R_2 \subseteq S_2} \cdots \sum_{R_p \subseteq S_p} f(R_1, R_2, \ldots, R_p) \quad (3.14)$$

to be the number of codewords which are always zero on the sets $N_i \setminus S_i$ (see Figure 3.3). It follows from the MDS property of the code that if only m symbols of an (n, k) MDS code are allowed to be nonzero, the $n - m$ zero symbols could be taken as information symbols, then the dimension of the resulting subcode is $k - n + m$ and

$$g(S_1, S_2, \ldots, S_p) = \begin{cases} 1, & \sum_{i=1}^{p} |S_i| < d \\ q^{1-d+\sum_{i=1}^{p} |S_i|}, & n \geq \sum_{i=1}^{p} |S_i| \geq d \end{cases}, \qquad (3.15)$$

Successively applying Möbius Inversion [109, Theorem 25.1] to (3.14), we get

$$f(R_1, R_2, \ldots, R_p) = \sum_{S_1 \subseteq R_1} \mu(S_1, R_1) \ldots \sum_{S_p \subseteq R_p} \mu(S_p, R_p) g(S_1, S_2, \ldots, S_p)$$

$$\triangleq \prod_{i=1}^{p} \left(\sum_{S_i \subseteq R_i} \mu(S_i, R_i) \right) g(S_1, S_2, \ldots, S_p), \qquad (3.16)$$

where

$$\mu(S, R) = \begin{cases} (-1)^{|R|-|S|}, & S \subseteq R \\ 0, & \text{otherwise} \end{cases}. \qquad (3.17)$$

Substituting (3.16) in (3.13),

$$A^T(w_1, w_2, \ldots, w_p) = \prod_{i=1}^{p-1} \left(\sum_{\substack{R_i \subseteq N_i \\ |R_i|=w_i}} \sum_{S_i \subseteq R_i} (-1)^{|R_i|-|S_i|} \right) G_p(\beta)$$

$$= \prod_{i=1}^{p-1} \left(\binom{n_i}{w_i} \sum_{j=0}^{w_i} \binom{w_i}{j} (-1)^{w_i-j} \right) G_p(\beta), \qquad (3.18)$$

such that $\beta = \sum_{i=1}^{p-1} |S_i|$ and by invoking (3.15)

$$G_p(\beta) = \sum_{\substack{R_p \subseteq N_p \\ |R_p|=w_p}} \sum_{S_p \subseteq R_p} (-1)^{|R_p|-|S_p|} g(S_1, S_2, \ldots, S_p)$$

$$= \binom{n_p}{w_p} \left(\sum_{i=0}^{d-\beta-1} \binom{w_p}{i} (-1)^{w_p-i} + \sum_{i=d-\beta}^{w_p} \binom{w_p}{i} (-1)^{w_p-i} q^{i+\beta-d+1} \right)$$

$$= \binom{n_p}{w_p} \sum_{i=d-\beta}^{w_p} \binom{w_p}{i} (-1)^{w_p-i} (q^{i+\beta-d+1} - 1). \qquad (3.19)$$

The last equality follows from the fact that $\sum_{j=0}^{w} \binom{w}{j}(-1)^{w-j} = (1-1)^w = 0$. Substituting (3.16) in (3.13), the theorem follows. $\qquad\square$

For the special case of two partitions, the split weight enumerator $A_{w_1,w_2}(n_1, n_2)$ is given in the following corollary.

Corollary 3.2. *Let $\mathcal{T}$ be an (n_1, n_2) partition of an (n, k, d) MDS code $\mathcal{C}$, then the split weight enumerator of $\mathcal{C}$ is*

$$A^{\mathcal{T}}(w_1, w_2) = \binom{n_1}{w_1}\binom{n_2}{w_2}\sum_{j=0}^{w_1}\binom{w_1}{j}(-1)^{w_1-j}\sum_{i=d-j}^{w_2}\binom{w_2}{i}(-1)^{w_2-i}(q^{i+j-d+1}-1).$$

From Theorem 3.1, it follows that the PWE of MDS codes does not depend on the orientation of the coordinates with respect to the partitions but only on the partitions' sizes and weights (see (3.14)). It thus intuitive that the ratio of $A^{\mathcal{T}}(w_1, w_2, \ldots, w_p)$ to $E(w)$ where $w = \sum_{i=1}^{p} w_i$ is the probability that the w nonzero symbols are distributed among the partitions with a $\mathcal{T}$-profile $(w_1, w_2, \ldots, w_p)$. Next we calculate this probability for the special case of $p = 2$ and we show that the partition weight enumerator admits to a simpler closed form formula.

Theorem 3.3. *Let $\mathcal{T}$ be an (n_1, n_2) partition for an (n, k, d) MDS code, $n = n_1 + n_2$, then*

$$A^{\mathcal{T}}(w_1, w_2) = E(w_1 + w_2)\frac{\binom{n_1}{w_1}\binom{n_2}{w_2}}{\binom{n}{w_1+w_2}}.$$

Proof. From Corollary 3.2, the split weight enumerator is

$$A^{\mathcal{T}}(w_1, w_2) =$$
$$\binom{n_1}{w_1}\binom{n_2}{w_2}\sum_{j=0}^{w_1}\binom{w_1}{j}(-1)^{w_1-j}\sum_{i=d-j}^{w_2}\binom{w_2}{i}(-1)^{w_2-i}(q^{i+j-d+1}-1). \quad (3.20)$$

Doing a change of variables, $\alpha = i + j$, we get

$$A^{\mathcal{T}}(w_1, w_2) =$$
$$\binom{n_1}{w_1}\binom{n_2}{w_2}\sum_{j=0}^{w_1}\binom{w_1}{j}(-1)^{w_1-j}\sum_{\alpha=\max(d,j)}^{w_2+j}\binom{w2}{\alpha-j}(-1)^{w_2-\alpha+j}(q^{\alpha-d+1}-1).$$

By changing the order of summation and summing over the same region:

$$A^T(w_1, w_2) =$$

$$\binom{n_1}{w_1}\binom{n_2}{w_2}\sum_{\alpha=d}^{w_1+w_2}(q^{\alpha-d+1}-1)(-1)^{w_1+w_2-\alpha}\sum_{j=0}^{\min(\alpha,w_1)}\binom{w_1}{j}\binom{w_2}{\alpha-j}$$

$$-\binom{n_1}{w_1}\binom{n_2}{w_2}\sum_{\alpha=w_2+1}^{w_1+w_2}(q^{\alpha-d+1}-1)(-1)^{w_1+w_2-\alpha}\sum_{j=0}^{\alpha-w_2-1}\binom{w_1}{j}\binom{w_2}{\alpha-j}.$$

By doing the change of variables $\beta = \alpha - w_2$ in the second summation

$$A^T(w_1, w_2) =$$

$$\binom{n_1}{w_1}\binom{n_2}{w_2}\sum_{\alpha=d}^{w_1+w_2}(q^{\alpha-d+1}-1)(-1)^{w_1+w_2-\alpha}\binom{w_1+w_2}{\alpha}$$

$$-\binom{n_1}{w_1}\binom{n_2}{w_2}\sum_{\beta=1}^{w_1}(q^{\alpha-d+1}-1)(-1)^{w_1+w_2-\alpha}\sum_{j=0}^{\beta-1}\binom{w_1}{j}\binom{w_2}{w_2+\beta-j}.$$

Since $\beta - j$ is always positive it follows that the second term in the right hand side is always zero and by letting $w = w_1 + w_2$

$$A^T(w_1, w_2) = \binom{n_1}{w_1}\binom{n_2}{w_2}\sum_{\alpha=d}^{w}\binom{w}{\alpha}(-1)^{w-\alpha}(q^{\alpha-d+1}-1). \tag{3.21}$$

By comparing with (3.10), the result follows. $\qquad\square$

Corollary 3.4. *The IOWE of a systematic MDS code, $O(w, h)$, for $h \geq d$, is given by*

$$O(w, h) = R(w, h-w) = E(h)\frac{\binom{k}{w}\binom{n-k}{h-w}}{\binom{n}{h}}$$

$$= \binom{k}{w}\binom{n-k}{h-w}\sum_{j=0}^{w}\binom{w}{j}(-1)^{w-j}\sum_{i=d-j}^{h-w}\binom{h-w}{i}(-1)^{h-w-i}(q^{i+j-d+1}-1).$$

By observing (3.3) and defining $\Psi(w)$ to be

$$\Psi(w) = \sum_{j=0}^{w} \binom{w}{j} (-1)^{w-j} \sum_{i=d-j}^{h-w} \binom{h-w}{i} (-1)^{h-w-i} (q^{i+j-d+1} - 1), \qquad (3.22)$$

we have an interesting identity:

$$\sum_{w=0}^{k} \Psi(w) \binom{k}{w} \binom{n-k}{h-w} = \Psi(0) \sum_{w=0}^{k} \binom{k}{w} \binom{n-k}{h-w}, \qquad (3.23)$$

where $\binom{n}{h} = \sum_{w=0}^{k} \binom{k}{w} \binom{n-k}{h-w}$ and $\Psi(0) = \sum_{i=d}^{h} \binom{h}{i} (-1)^{h-i} (q^{i-d+1} - 1)$.

Corollary 3.5. *For an (n,k,d) MDS code $\mathcal{C}$, the number of codewords which are exactly nonzero at a fixed subset of coordinates of cardinality h and are zero at the remaining h coordinates is $\frac{E(h)}{\binom{n}{h}}$.*

Proof. Let $\mathcal{T}$ be the implied $(h, n-h)$ partition, then the required number of codewords is $A^{\mathcal{T}}(h, 0)$. The result follows by applying Theorem 3.3. □

This result illustrates how the partition weight enumerator of MDS codes is independent of the orientation of the partitions. Since there are $E(h)$ codewords of weight h and there are $\binom{n}{h}$ distinct ways to choose the h zero coordinates, then in such a case one expects that that there are $\frac{E(h)}{\binom{n}{h}}$ codewords for any choice of the h coordinates.

By following the same lines of proof, the result of Theorem 3.3 can be generalized to an arbitrary number of partitions as in the following theorem:

Theorem 3.6. *For an (n,k,d) MDS code $\mathcal{C}$ with an $(n_1, n_2, \ldots, n_p)$ partition of its coordinates the p-partition weight enumerator is given by*

$$A^{\mathcal{T}}(w_1, w_2, \ldots, w_p) = E(w) \frac{\binom{n_1}{w_1} \binom{n_2}{w_2} \cdots \binom{n_p}{w_p}}{\binom{n}{w}},$$

where $w = \sum_{i=1}^{p} w_i$ and $E(w) = |\{\mathbf{c} \in \mathcal{C} : \mathcal{W}(\mathbf{c}) = w\}|$.

We give numerical examples of PWEs using Theorem 3.1 and Theorem 3.6. For these examples, the PWGFs were also verified numerically by generating the $(7, 3, 5)$ RS code.

Example 3.1. The PWGF for the $(1,1,2,3)$ partition of the coordinates of the $(7,3,5)$ RS code over F_8 is

$$
\begin{aligned}
\mathbb{P}(\mathcal{V},\mathcal{X},\mathcal{Y},\mathcal{Z}) = & 1 + 21\mathcal{V}\mathcal{X}\mathcal{Y}^2\mathcal{Z} + 42\mathcal{V}\mathcal{X}\mathcal{Y}\mathcal{Z}^2 + 21\mathcal{V}\mathcal{Y}^2\mathcal{Z}^2 + 21\mathcal{X}\mathcal{Y}^2\mathcal{Z}^2 + 63\mathcal{V}\mathcal{X}\mathcal{Y}^2\mathcal{Z}^2 \\
& + 7\mathcal{V}\mathcal{X}\mathcal{Z}^3 + 14\mathcal{V}\mathcal{Y}\mathcal{Z}^3 + 14\mathcal{X}\mathcal{Y}\mathcal{Z}^3 + 42\mathcal{V}\mathcal{X}\mathcal{Y}\mathcal{Z}^3 + 7\mathcal{Y}^2\mathcal{Z}^3 + 21\mathcal{V}\mathcal{Y}^2\mathcal{Z}^3 \\
& + 21\mathcal{X}\mathcal{Y}^2\mathcal{Z}^3 + 217\mathcal{V}\mathcal{X}\mathcal{Y}^2\mathcal{Z}^3.
\end{aligned}
$$

It can be checked that the sum of the coefficients is the total number of codewords 8^3. For this example, one can also verify the PWGF numerically.   ◇

Example 3.2. The $(3,2,2)$ 3-partition enumerator of the $(7,5,3)$ RS code over F_8 is

$$
\begin{aligned}
\mathbb{P}(\mathcal{X},\mathcal{Y},\mathcal{Z}) = & 1 + 7\mathcal{X}^3 + 42\mathcal{X}^2\mathcal{Y} + 70\mathcal{X}^3\mathcal{Y} + 21\mathcal{X}\mathcal{Y}^2 + 105\mathcal{X}^2\mathcal{Y}^2 + 266\mathcal{X}^3\mathcal{Y}^2 \\
& + 42\mathcal{X}^2\mathcal{Z} + 70\mathcal{X}^3\mathcal{Z} + 84\mathcal{X}\mathcal{Y}\mathcal{Z} + 420\mathcal{X}^2\mathcal{Y}\mathcal{Z} + 1064\mathcal{X}^3\mathcal{Y}\mathcal{Z} + 14\mathcal{Y}^2\mathcal{Z} \\
& + 210\mathcal{X}\mathcal{Y}^2\mathcal{Z} + 1596\mathcal{X}^2\mathcal{Y}^2\mathcal{Z} + 3668\mathcal{X}^3\mathcal{Y}^2\mathcal{Z} + 21\mathcal{X}\mathcal{Z}^2 + 105\mathcal{X}^2\mathcal{Z}^2 \\
& + 266\mathcal{X}^3\mathcal{Z}^2 + 14\mathcal{Y}\mathcal{Z}^2 + 210\mathcal{X}\mathcal{Y}\mathcal{Z}^2 + 1596\mathcal{X}^2\mathcal{Y}\mathcal{Z}^2 + 3668\mathcal{X}^3\mathcal{Y}\mathcal{Z}^2 \\
& + 35\mathcal{Y}^2\mathcal{Z}^2 + 798\mathcal{X}\mathcal{Y}^2\mathcal{Z}^2 + 5502\mathcal{X}^2\mathcal{Y}^2\mathcal{Z}^2 + 12873\mathcal{X}^3\mathcal{Y}^2\mathcal{Z}^2.
\end{aligned}
$$

It can also be verified that $\mathbb{P}(1,1,1) = 8^3$.   ◇

Theorem 3.6 implies that the distribution of the $wE(w)$ nonzero symbols within the codewords of the same Hamming weight w is uniform among the partitions. This issue will be addressed in more detail in the following section.

3.3   A Relationship Between Coordinate Weight and Codeword Weight

In this section, we will show that for MDS codes, one can derive the coordinate weight from the codeword weight. We will discuss whether other linear codes also have this property.

Define $\mathcal{C}_h$ to be the subcode of $\mathcal{C}$ with codewords of Hamming weight h;

$$
\mathcal{C}_h \triangleq \{\boldsymbol{c} \in \mathcal{C} : \mathcal{W}(\boldsymbol{c}) = h\}. \tag{3.24}
$$

The following lemma calculates the total weight of any coordinate in the set $\mathcal{C}_h$.

Lemma 3.7. *For an (n, k, d) MDS code $\mathcal{C}$ the total Hamming weight of any coordinate, summed over the subcode $\mathcal{C}_h$, is equal to $\frac{h}{n}E(h)$.*

Proof. Let $\mathcal{T}$ be an $(1, n-1)$ partition of $\mathcal{C}$, where the coordinate of choice forms the partition of size one. By Theorem 3.3, it follows that for any such partition, the number of codewords of $\mathcal{C}$ which are nonzero in this coordinate and have a total weight h, i.e., a weight profile $(1, h-1)$, is

$$A^{\mathcal{T}}(1, h-1) = \frac{\binom{n-1}{h-1}}{\binom{n}{h}}E(h) = \frac{h}{n}E(h). \tag{3.25}$$

By observing that $A^{\mathcal{T}}(1, h-1)$ is the total weight of the chosen coordinate over codewords in $\mathcal{C}_h$ and that the choice of that coordinate was arbitrary, we are done. □

This means that the codewords of the subcode $\mathcal{C}_h$, when arranged as the rows of an array, result in a design where the Hamming weight of each row is h and the Hamming weight of each column is $\frac{h}{n}E(h)$. Furthermore, the Hamming distance between any two rows is at least $d = n - k + 1$. We are now ready to prove an important property of MDS codes:

Theorem 3.8. *For an (n, k, d) MDS code $\mathcal{C}$, the ratio of the total weight of any s coordinates of $\mathcal{C}_h$ to the total weight of $\mathcal{C}_h$ is $\frac{s}{n}$. If the s coordinates are "input" coordinates, then*

$$\frac{\sum_{w=1}^{s} w\, O(w, h)}{s} = \frac{h\, E(h)}{n}$$

for any Hamming weight h.

Proof. By Lemma 3.7, the total weight of any coordinate of $\mathcal{C}_h$ is $(h/n)E(h)$. The total weight of any s coordinates of $\mathcal{C}_h$ is the sum of the weights of the individual coordinates, $s(h/n)E(h)$. By observing that the weight of the s coordinates can be also expressed in terms of the IOWE by $\sum_{w=1}^{s} wO(w, h)$ and $hE(h)$ is the total weight of $\mathcal{C}_h$, the theorem follows. □

As a side result, we have proven this identity (c.f., (3.23)):

Corollary 3.9. *Let* $\Psi(w)$ *be defined as in (3.22) then*

$$\sum_w \Psi(w) \binom{s-1}{w-1}\binom{n-s}{h-w} = \Psi(0) \sum_w \binom{s-1}{w-1}\binom{n-s}{h-w}.$$

Proof. For an $\mathcal{T} : (s, n-s)$ partition of the coordinates, it follows from Theorem 3.8 that $\sum_{w=1}^{s} \frac{w}{s} A^T(w, h - w) = \frac{h}{n}E(h) = \binom{n-1}{h-1}\Psi(0)$. Also by Corollary 3.2, $\sum_{w=1}^{s} \frac{w}{s} A^T(w, h - w) = \sum_{w=1}^{s} \binom{s-1}{w-1}\binom{n-s}{h-w}\Psi(w)$. The proof follows from the identity $\binom{n-1}{h-1} = \sum_w \binom{s-1}{w-1}\binom{n-s}{h-w}$.

$\square$

Definition 3.1. An (n, k) code $\mathcal{C}$ (not necessary MDS) is said to have the multiplicity property $\mathcal{M}$, if for any $\mathcal{T} : (s, n - s)$ partition, $\sum_{w=1}^{s} \frac{w}{s} A^T(w, h - w) = \frac{h}{n}E(h)$ for all Hamming weights h.

We will refer to the partition composed of the s coordinates as the input partition. By Theorem 3.8, all MDS codes have property $\mathcal{M}$. In general not all linear codes have property $\mathcal{M}$ as seen in the following counterexample:

Example 3.3. The $(5, 3)$ linear code defined by

$$G = \begin{pmatrix} 1 & 0 & 0 & 1 & 1 \\ 0 & 1 & 0 & 0 & 1 \\ 0 & 0 & 1 & 0 & 1 \end{pmatrix}$$

is composed of the 8 codewords $00000, 10011, 01001, 11010, 00101, 10110, 01100, 11111$. Let the input partition be composed of the first 3 coordinates. For $s = k = 3$, let $\beta(h) = \sum_w wO(w, h)$; and $\xi(h) = \frac{3}{5}hE(h)$, then from the following table it is clear that it is not true that this code has property $\mathcal{M}$.

h:	0	1	2	3	4	5
$\beta(h)$:	0	0	4	5	0	3
$\xi(h)$:	0	0	3.6	5.4	0	3

$\diamond$

It is to be noted that all cyclic codes have property $\mathcal{M}$. This is partially justified by the fact that any cyclic shift of a codeword of weight h is also a codeword of weight

h with h/n of the coordinates holding nonzero elements [107]. However, this neither implies Theorem 3.8 nor is it implied by Theorem 3.8. For example, an extended RS code is an MDS code but not a cyclic code while an $(7,4)$ binary Hamming code is cyclic but not MDS. Also, if a code satisfies property $\mathcal{M}$, it is not necessary that the code is either cyclic or MDS. For example, the first-order Reed-Muller codes as well as their dual codes, the extended Hamming codes, have property $\mathcal{M}$ but are neither cyclic nor MDS. Next, we discuss some codes with the multiplicity property.

Theorem 3.10. *The first-order Reed-Muller codes have the multiplicity property $\mathcal{M}$.*

Proof. The weight enumerator of the first-order Reed-Muller codes of length 2^m, $\mathcal{R}(1,m)$, is $\mathbb{E}(\mathcal{W}) = 1 + (2^{m+1} - 2)\mathcal{W}^{2^{m-1}} + \mathcal{W}^{2^m}$ and their minimum distance is 2^{m-1}. Let H_{2^m} be the Hadamard matrix of order 2^m and let M be the binary matrix that results from stacking H_{2^m} on top $-H_{2^m}$ and replacing each $+1$ by 0 and each -1 by 1. (A Hadamard matrix H of order n is an $n \times n$ matrix with entries $+1$ and -1 such that $HH^T = nI$ and I is the identity matrix. [109, Chapter 18].) The codewords of $\mathcal{R}(1,m)$ are exactly the rows of M [109, Chapter 18]. It follows that each codeword of weight 2^{m-1} has a unique codeword of the same weight which is its binary complement. Thus each coordinate will be equally one and zero in half the number of such codewords. Since the remaining codewords are the all-zero and the all-one codewords, it follows that $\mathcal{R}(1,m)$ has the multiplicity property. $\qquad\square$

We now prove here that if a linear code has property $\mathcal{M}$ then its dual code also has property $\mathcal{M}$. By a straightforward manipulation of the McWilliams identities [74, Chapter 5, (52)] one can show the following relationship between the PWEs of a code and its dual code [26] (c.f., Theorem 6.5):

Theorem 3.11. *Let $\mathcal{C}$ be an (n,k) linear code over $\mathbb{F}_q$ and $\mathcal{C}^\perp$ be its dual code. If $\mathcal{T}$ is an $(n1, n2)$ partition of their coordinates, $A(\alpha, \beta)$ and $A^\perp(\alpha, \beta)$ are the PWEs of $\mathcal{C}$ and $\mathcal{C}^\perp$ respectively, then $A(\alpha, \beta)$ and $A^\perp(\alpha, \beta)$ are related by*

$$A^\perp(\alpha, \beta) = \frac{1}{|\mathcal{C}|} \sum_{v=0}^{n_2} \sum_{w=0}^{n_1} A(w,v)\mathcal{K}_\alpha(w, n_1)\mathcal{K}_\beta(v, n_2),$$

such that the Krawtchouk polynomial is $\mathcal{K}_\beta(v,\gamma) = \sum_{j=0}^{\beta} \binom{\gamma-v}{\beta-j}\binom{v}{j}(-1)^j(q-1)^{\beta-j}$ *for* $\beta = 0, 1, \dots, \gamma$.

Define $A_i(\alpha, \beta)$ and $A_i^{\perp}(\alpha, \beta)$ to be the PWEs for $\mathcal{C}$ and $\mathcal{C}^{\perp}$ respectively when an $(1, n-1)$ partition is applied to their coordinates such that the first partition of cardinality one is composed of the ith coordinate.

Theorem 3.12. *An (n, k) linear code over $\mathbb{F}_q$ has the multiplicity property iff its dual code has the multiplicity property.*

Proof. Let $\mathcal{C}$ be an (n, k) linear code over $\mathbb{F}_q$ with property $\mathcal{M}$ and an $(1, n-1)$ PWE $A_i(\alpha, \beta)$. From Theorem 3.11 the PWE of the dual code $\mathcal{C}^{\perp}$ is

$$A_i^{\perp}(1, \beta) = \frac{1}{|\mathcal{C}|} \sum_{v=0}^{n-1} \sum_{w=0}^{1} A_i(w, v)\mathcal{K}_1(w, 1)\mathcal{K}_\beta(v, n-1). \qquad (3.26)$$

Since $\mathcal{C}$ has property $\mathcal{M}$, then $A_i(1, v) = \frac{v+1}{n}E_{\mathcal{C}}(v+1)$ and $A_i(0, v) = E_{\mathcal{C}}(v) - A_i(1, v-1) = (1 - \frac{v}{n})E_{\mathcal{C}}(v)$. By substituting in (3.26), it follows that $A_i^{\perp}(1, \beta) = A_j^{\perp}(1, \beta)$ for any $i, j \in \{1, 2, \dots, n\}$ and $\sum_{i=1}^{n} A_i^{\perp}(1, \beta) = nA_i^{\perp}(1, \beta)$ for any i. Counting the total weight of the codewords in $\mathcal{C}^{\perp}$ with Hamming weight h by two different ways, we get $\sum_{i=1}^{n} A_i^{\perp}(1, \beta) = (\beta+1)E_{\mathcal{C}^{\perp}}(\beta+1)$. It follows that $A_i^{\perp}(1, \beta) = \frac{\beta+1}{n}E_{\mathcal{C}^{\perp}}(\beta+1)$ and $\mathcal{C}^{\perp}$ has property $\mathcal{M}$.

For the converse, assume that $\mathcal{C}$ does not satisfy property $\mathcal{M}$ but $\mathcal{C}^{\perp}$ does. From the previous argument $(\mathcal{C}^{\perp})^{\perp}$ has property $\mathcal{M}$. Since for linear codes $(\mathcal{C}^{\perp})^{\perp} = \mathcal{C}$, we reach a contradiction. □

Since the dual codes of MDS codes are also MDS codes, this result strengthens Theorem 3.8. This theorem somehow strengthens the result of Theorem 3.8 since the dual codes of MDS codes are again MDS codes. The dual codes of cyclic codes are also cyclic codes. One can also use this theorem to show that certain codes have the multiplicity property.

Corollary 3.13. *The extended Hamming codes have property $\mathcal{M}$.*

Proof. An extended Hamming code of length 2^m is the dual of the first-order RM code $\mathcal{R}(1, m)$ [74], which by Theorem 3.10 has property $\mathcal{M}$. □

It is also the case that if the code has a transitive automorphism group then the code has the multiplicity property [19]. Extended Hamming codes also have transitive automorphism groups [19] which gives another proof to Corollary 3.13. Some product codes also have the multiplicity property [19, 27].

3.4 Binary Partition Weight Enumerator of MDS Codes

In this section, we study the partition weight enumerator of the binary image of an RS (MDS) code. Let $\mathcal{T}$ be a partition of the coordinates of an MDS code $\mathcal{C}$ defined over $\mathbb{F}_{2^m}$. Let $\mathcal{T}_b$ be the partition of the coordinates of the code's binary image $\mathcal{C}^b$ implied by $\mathcal{T}$ when each symbol is represented with its binary image. The number of the partitions in $\mathcal{T}$ and $\mathcal{T}_b$ is the same but the size of each partition is m times larger. This is illustrated by example in Figure 3.4. The *binary partition weight enumerator* (PWE) gives the number of codewords in the binary image with a specific combination of binary Hamming weights in the specified partitions. As we saw in the Section 2.2, the binary image is not unique, so we will resort again to an *averaged* binary PWE.

Theorem 3.14. *Let $\mathbb{P}^{\mathcal{T}}(\mathcal{X}_1, \mathcal{X}_2, \ldots, \mathcal{X}_p)$ be the partition weight generating function (PWGF) of an (n, k) code over F_{2^m}, and $\mathcal{T}_b$ be the partitioning of the coordinates of $\mathcal{C}^b$ induced by $\mathcal{T}$ when the symbols in each partition are represented by bits, then the average binary PWGF is*

$$\tilde{\mathbb{P}}^{\mathcal{T}_b}_{\mathcal{C}^b}(\mathcal{Z}_1, \mathcal{Z}_2, \ldots, \mathcal{Z}_p) = \mathbb{P}^{\mathcal{T}}_{\mathcal{C}}(F(\mathcal{Z}_1), F(\mathcal{Z}_2), \ldots, F(\mathcal{Z}_p)),$$

where $F(\mathcal{Z}) = \frac{1}{2^m-1}(1+\mathcal{Z})^m - 1$.

Proof. Assuming a binomial distribution of the bits in a nonzero symbol, the probability that the binary representation of a nonzero symbol has weight i is equal to the coefficient of $\mathcal{Z}^i$ in $\frac{1}{2^m-1}\sum_{i=1}^{m} \binom{m}{i}\mathcal{Z}^i$. If the weight of the jth partition is w_j, then the average binary weight generator function of its binary image is $\left(\frac{1}{2^m-1}\sum_{i=1}^{m}\binom{m}{i}\mathcal{Z}_j^i\right)^{w_j}$ under the assumption that all the nonzero symbols are independent and equally probable. Consider a codeword with a weight profile $(w_1, w_2, \ldots, w_p)$, then the probability

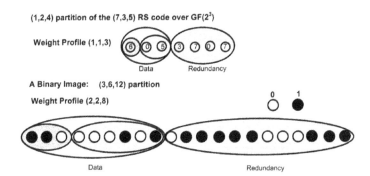

Figure 3.4: Partitioning of a code and its binary image.
A codeword in the $(7,3,5)$ RS code over $\mathbb{F}_8$ is shown with a $(1,2,4)$ partition of its coordinates. For a specific binary representation, the binary image is shown with the implied $(3,6,12)$ partition of its coordinates. We emphasize that the weight profile of the binary image is not easily derived from that on the symbol level.

that the weight profile of its binary image is $(b_1, b_2, \ldots, b_p)$ is given by the coefficient of $\mathcal{Z}_1^{b_1} \mathcal{Z}_2^{b_2} \ldots \mathcal{Z}_p^{b_p}$ in $\prod_{j=1}^{p} \left(\frac{1}{2^m-1} \sum_{i=1}^{m} \binom{m}{i} \mathcal{Z}_j^i \right)^{w_j}$. By multiplying with the number of such codewords, $A^T(w_1, w_2, \ldots, w_p)$, the result follows. $\qquad \square$

For systematic codes, the binary IOWE could be derived from the binary PWE as in (3.8) (Unless otherwise stated, when speaking of binary weight enumerators of codes over F_{2^m} it is understood that we mean the ensemble average binary weight enumerator.) For example, the coefficient of $\mathcal{X}^w \mathcal{Y}^h$ in $\tilde{\mathbb{P}}(\mathcal{X}\mathcal{Y}, \mathcal{Y}, \ldots, \mathcal{Y})$ is the number of codewords with input binary weight w in the first partition and a total average binary weight h. In the following corollary, we give a closed form expression for the binary IOWE, $\tilde{O}(w_b, h_b)$.

Corollary 3.15. *Let $O_{\mathcal{C}}(w, h)$ be the input-output weight enumerator of an (n, k, d) code $\mathcal{C}$, defined over $\mathbb{F}_{2^m}$ corresponding to an $(s, n-s)$ partition of its coordinates, then* *the average binary IOWE of $\mathcal{C}^b$ is given by*

$$\tilde{O}_{\mathcal{C}^b}(w_b, h_b) = \sum_{w=0}^{s} \sum_{h=w}^{n} \frac{O_{\mathcal{C}}(w, h)}{(2^m - 1)^h}$$

$$\left(\sum_{j=0}^{h-w} (-1)^{h-w-j} \binom{h-w}{j} \binom{jm}{h_b - w_b} \right) \left(\sum_{j=0}^{w} (-1)^{w-j} \binom{w}{j} \binom{jm}{w_b} \right)$$

for $h_b \geq d$.

Proof. For the given $(s, n-s)$ partition, the split weight enumerator of $\mathcal{C}$ is $\mathbb{P}_{\mathcal{C}}(\mathcal{X}, \mathcal{Y}) = \sum_{w=0}^{s} \sum_{h=w}^{n} O_{\mathcal{C}}(w, h) \mathcal{X}^w \mathcal{Y}^{h-w}$. From the Theorem 3.14 and (3.6), $\tilde{O}_{\mathcal{C}^b}(w_b, h_b)$ is the coefficient of $\mathcal{X}^{w_b} \mathcal{Y}^{h_b}$ in

$$\tilde{\mathbb{O}}_{\mathcal{C}^b}(\mathcal{X}, \mathcal{Y}) = \frac{1}{(2^m - 1)^h} \sum_{w=0}^{s} \sum_{h=w}^{n} O_{\mathcal{C}}(w, h)((1 + \mathcal{Y}\mathcal{X})^m - 1)^w ((1 + \mathcal{Y})^m - 1)^{h-w}. \quad (3.27)$$

Since $((1 + \mathcal{Y}\mathcal{X})^m - 1)^w = \sum_{j=0}^{w} \binom{w}{j}(-1)^{w-j}(\sum_{i=0}^{mj} \binom{mj}{i} \mathcal{X}^i \mathcal{Y}^i)$ and $((1 + \mathcal{Y})^m - 1)^{h-w} = \sum_{j=0}^{h-w} \binom{h-w}{j}(-1)^{h-w-j}(\sum_{i=0}^{mj} \binom{mj}{i} \mathcal{Y}^i)$, the result follows by substituting in (3.27). $\quad\square$

The IOWE of the binary image will be useful in the analysis of the bit error probability of MDS codes when their binary image is transmitted. In Section 3.3 (c.f., Theorem 3.8), we showed that MDS codes have the multiplicity property. Now, we will show that ensemble binary image of an MDS code will also have the multiplicity property on average.

Theorem 3.16. *Let $\mathcal{C}$ be an (n, k, d) MDS code over $\mathbb{F}_{2^m}$ with the multiplicity property and $\tilde{E}(h_b)$ be the average binary weight enumerator of $\mathcal{C}^b$. If $\tilde{O}(w_b, h_b)$ is the average binary IOWE of $\mathcal{C}^b$, where the partition of the coordinates of $\mathcal{C}^b$ is induced by an $(s, n-s)$ partition of the coordinates of $\mathcal{C}$, then for $h_b \geq d$*

$$\frac{\sum_{w_b=1}^{ms} w_b \, \tilde{O}(w_b, h_b)}{m \, s} = \frac{h_b \, \tilde{E}(h_b)}{m \, n}.$$

Proof. We will begin by proving it for the special case of $s = 1$. Since $\mathcal{C}$ has property

$\mathcal{M}$, then $O(1, h) = \frac{h}{n} E(h)$. It follows from Corollary 3.15 that

$$\tilde{O}(w_b, h_b) = \binom{m}{w_b} \sum_{h=0}^{n} \frac{h}{n} \frac{E(h)}{(2^m - 1)^h} \sum_{j=0}^{h-1} (-1)^{h-1-j} \binom{h-1}{j} \binom{jm}{h_b - w_b}. \tag{3.28}$$

By changing the order of the summations we have

$$\sum_{w_b=1}^{m} w_b \tilde{O}(w_b, h_b) = \sum_{h=0}^{n} \frac{h}{n} \frac{E(h)}{(2^m - 1)^h} \sum_{j=0}^{h-1} (-1)^{h-1-j} \binom{h-1}{j} \sum_{w_b=1}^{m} w_b \binom{m}{w_b} \binom{jm}{h_b - w_b}. \tag{3.29}$$

By observing that $w_b \binom{m}{w_b} = m \binom{m-1}{w_b-1}$, it follows that the rightmost summation in (3.30) is equal to $m \sum_{w_b} \binom{m-1}{w_b-1} \binom{mj}{h_b-1-(w_b-1)} = m \binom{m(j+1)-1}{h_b-1}$. By doing a change of variables $\alpha = j+1$ and observing that $\binom{m\alpha-1}{h_b-1} = \frac{h_b}{m\alpha} \binom{m\alpha}{h_b}$ and rearranging, it follows that the total weight of m coordinates in the binary image $\mathcal{C}_b$, corresponding to a single coordinate in $\mathcal{C}$, is

$$\sum_{w_b=1}^{m} w_b \tilde{O}(w_b, h_b) = \frac{1}{n} h_b \sum_{h=1}^{n} \frac{E(h)}{(2^m - 1)^h} \sum_{\alpha=1}^{h} (-1)^{h-\alpha} \binom{h}{\alpha} \binom{m\alpha}{h_b}$$

$$= \frac{h_b}{n} \tilde{E}(h_b). \tag{3.30}$$

If the input partition has s coordinates of $\mathcal{C}$, the result follows by summing the weights of the individual coordinates.   □

This means that if the weight of a symbol coordinate is $(h/n)E(h)$ in $\mathcal{C}_h$, then the average weight of its binary image is $(h_b/n)\tilde{E}(h_b)$ in $\mathcal{C}_{h_b}^b$. It will be interesting to determine whether this will still be true for any binary representation. As we will see in the next section, the result of Theorem 3.16 can simplify the analysis of the bit error probability of MDS codes.

3.5   Symbol and Bit Error Probabilities

In Section 2.4, we showed how one can analyze the codeword error probability of various RS code decoders. In this section, we study the symbol and bit error probabilities of systematic MDS codes. In general, systematic coding is preferred over nonsystematic coding. It has also been shown that maximum-likelihood (ML) decoding of binary linear codes achieves the least bit error probability when the code is systematic [43].

Given a symbol-level decoder (soft-decision or hard-decision decoder), the codeword error error probability (CEP) at a certain signal-to-noise ratio (SNR) γ will be a function of the SNR γ and the code weight enumerator $E(h)$. In the remaining of this chapter, we will denote the CEP at a signal-to-noise ratio (SNR) γ by $\Phi_c(E(h), \gamma)$. For linear codes, union upper-bounds on the performance of symbol-based decoders are of the form

$$\Phi_c(E(h), \gamma) \leq \sum_{h=d}^{n} E(h)\mathcal{U}(\gamma, h), \tag{3.31}$$

for some function $\mathcal{U}$ of the SNR γ and weight h.

Tighter upper bounds can be of the form

$$\Phi_c(E(h), \gamma) \leq \min_{\alpha} \left\{ \sum_{h=d}^{\alpha} E(h)\mathcal{V}(\gamma, h) + \mathcal{F}(\gamma, \alpha) \right\}, \tag{3.32}$$

for some functions $\mathcal{V}$ and $\mathcal{F}$ of γ and h. For example, tight upper bounds on the performance of bit-level and symbol-level hard-decision maximum-likelihood decoders admit to the above form and are given by (8.32) and Theorem 8.9 respectively. The codeword error probability of the HD Berlekamp-Massey decoder is the probability that the received word lies in the decoding sphere of a codeword other than the transmitted word. It is also determined by the weight enumerator and has the form of the union bound as in (3.31);

$$\Phi_c(E(h), \gamma) \leq \sum_{h=d}^{n} E(h) \sum_{t=0}^{\tau} P_t^h(\gamma), \tag{3.33}$$

where $P_t^h(\gamma)$ is the probability that a received word is exactly Hamming distance t from a codeword of weight h and $\tau = \lfloor (d-1)/2 \rfloor$ is the Hamming decoding radius [79] [115].

Given an upper bound on the CEP of a symbol-based decoder, it is well known that the symbol error probability (SEP) $\Phi_s(\gamma)$ can be derived from the CEP $\Phi_c(\gamma)$ by substituting $E(h)$ with

$$Q(k,h) = \sum_{w=1}^{k} \frac{w}{k} O(w,h), \tag{3.34}$$

(e.g., [115, (10-14)]). From Theorem 3.8, the common approximation

$$Q(k,h) \approx \frac{h}{n} E(h) \tag{3.35}$$

is exact for MDS codes and

$$\Phi_s(\gamma) = \Phi_c\left(E(h),\gamma\right)\Big|_{E(h):=Q(k,h)}. \tag{3.36}$$

In other words, if the CEP is given by (3.31) or (3.32), the SEP will be respectively bounded by

$$\Phi_s(\gamma) \leq \sum_{h=d}^{n} \frac{h}{n} E(h)\mathcal{U}(\gamma,h), \tag{3.37}$$

$$\Phi_s(\gamma) \leq \min_{\alpha}\left\{\sum_{h=d}^{\alpha} \frac{h}{n} E(h)\mathcal{V}(\gamma,h) + \mathcal{F}(\gamma,\alpha)\right\}. \tag{3.38}$$

In case the binary image of an RS code is transmitted and the decoder is a bit-level decoder, performance analysis of the decoder will utilize the binary weight enumerator of the code. As we discussed in Section 2.4, the ensemble average binary weight enumerators become handy when analyzing the performance of the binary images of RS codes. As is the case of symbol based decoders, upper bound on the CEP of bit-level decoders admit the union bound forms

$$\Phi_c\left(\tilde{E}(h),\gamma\right) \leq \sum_{h=d}^{nm} \tilde{E}(h)\Upsilon(\gamma,h) \tag{3.39}$$

$$\Phi_c\left(\tilde{E}(h),\gamma\right) \leq \min_{\alpha}\left\{\sum_{h=d}^{\alpha} \tilde{E}(h)\mathcal{J}(\gamma,h) + \mathcal{G}(\gamma,\alpha)\right\} \tag{3.40}$$

for some functions Υ, $\mathcal{J}$ and $\mathcal{G}$ of the SNR γ and the weight h. For example, the union bounds of SD and HD decoding of (2.36) and (2.39) are of the form of (3.39), whereas the Poltyrev tighter version of these bounds follow the form of (3.40).

From Theorem 3.16, we know that for any k (symbol) coordinates of the MDS code

$$\tilde{Q}(mk,h) = \sum_{w=1}^{mk} \frac{w}{mk}\tilde{O}(w,h) = \frac{h}{mn}\tilde{E}(h). \tag{3.41}$$

It follows that the bit error probability (BEP) can be bounded by (e.g., [8, 98])

$$\Phi_b(\gamma) = \Phi_c\left(\tilde{E}(h),\gamma\right)\Big|_{\tilde{E}(h):=\tilde{Q}(mk,h)} \tag{3.42}$$

$$\leq \min_\alpha\left\{\sum_{h=d}^\alpha \frac{h}{mn}\tilde{E}(h)\mathcal{J}(\gamma,h) + \mathcal{G}(\gamma,\alpha)\right\} \tag{3.43}$$

$$\leq \sum_{h=d}^{nm} \frac{h}{mn}\tilde{E}(h)\Upsilon(\gamma,h). \tag{3.44}$$

3.6 Multiuser Error Probability

We consider the case when a systematic RS code is shared among different users or applications. The systematic symbols are shared among the different users where the coordinates of the code are partitioned according to an $\mathcal{T} : (n_1, n_2, ..., n_{p-1}, n-k)$ partition. The ith partition of size n_i is assigned to the ith user and the last partition constitutes of the redundancy symbols. Since the considered codes are linear, we assume that the all zero codeword is transmitted. If a codeword of symbol weight h and of weight w_j in the jth partition is erroneously decoded, a fraction $\frac{w_j}{n_j}$ of the jth user's symbols are received in error. It follows that the jth user's symbol error probability could be written as (c.f., (3.49))

$$\Phi_s^j(\gamma) = \Phi_c\left(Q^j(n_j,h),\gamma\right), \tag{3.45}$$

where

$$Q^j(n_j,h) = \sum_{w=1}^{n_j} \frac{w}{n_j}O^j(w,h) \tag{3.46}$$

and $O^j(w,h)$ is the jth partition input-output weight enumerator derived from the PWE as in (3.8). The following theorem gives an important result regarding the multiuser error probability of MDS (RS) codes:

Theorem 3.17. *If a systematic linear MDS code is shared among different users, all users have the same unconditional symbol error probability regardless of the sizes of the partitions assigned to them.*

Proof. The SEP of a certain user j, whose partition's size is n_j, is given by (3.45). Thus, it is sufficient to show that for two different users i and j with partitions of sizes n_i and n_j respectively, such that $n_i \neq n_j$, $Q^j(n_j, h) = Q^i(n_i, h)$. From Theorem 3.8, it follows that for an arbitrary partition of size n_j, $Q^j(n_j, h) = \frac{h}{n}E(h)$. Since this result does not depend on the size of the partition nor on the orientation of the coordinates with respect to it, we are done. $\qquad\qquad\square$

Now, consider the case when the binary image of an RS code is transmitted and the decoder is a bit-level hard-decision or soft-decision decoder. The systematic coordinates will be partitioned among different users where the partitions on the bit level will follow from the partitions on the symbol level (e.g., Figure 3.4). In case of a bit-level decoder, the bit error probability of the jth user can be given by

$$\Phi_b^j(\gamma) = \Phi_c\left(\tilde{Q}^j(mn_j, h), \gamma\right), \qquad (3.47)$$

such that

$$\tilde{Q}^j(mn_j, h) = \sum_{w=1}^{mn_j} \frac{w}{mn_j}\tilde{O}^j(w, h), \qquad (3.48)$$

where $\tilde{O}^j(w, h)$ is the average binary input-output weight enumerator of the jth user and $\frac{w}{mn_j}\tilde{O}^j(w, h)$ is the fraction of the jth user's bits received in error when a codeword of total weight h and weight w in the jth partition is erroneously decoded given that the all zero codeword was transmitted.

Theorem 3.18. *For systematic MDS linear codes, the average unconditional bit error probability of all users is the same regardless of the number of symbols in each partition or the orientation of the partition assigned to them.*

Proof. Let users i and j be assigned two different partitions of $\mathcal{C}$ with different sizes

n_i and n_j. Now consider the binary images of these partitions. Equations (3.41) and (3.47) imply that both users have the same average bit error probability. □

Now that we have shown that the unconditional symbol and bit error probability are the same for all partitions (users) regardless of their size, we can ask questions about the conditional error probability. Using the results in this chapter, one could answer interesting questions about the conditional multiuser error probability. Since the code is linear, we will assume that the all-zero codeword is transmitted. For example, the conditional CEP given that for any codeword no more than a fraction p of the jth user's symbols are ever received in error is given by [2]

$$\underline{\Phi_c}(\gamma) \;\; = \;\; \Phi_c \left(\sum_{w_j=0}^{\lfloor pm_j \rfloor} O^j(w_j, h), \gamma \right) \tag{3.49}$$

where a hard-decision symbol level decoder with a decoding radius τ was assumed. We only considered error events due to codewords whose weight in the jth partition is not greater than pn_j. Recall that in the unconditional case $\sum_{w_j=0}^{\lfloor pm_j \rfloor} O^j(w_j, h)$ is replaced by $E(h) = \sum_{w_j=0}^{n_j} O^j(w_j, h)$.

Define the following weight enumerator

$$O^{i,j}(w_i, w_j, h) \triangleq |\{c \in \mathcal{C} : (\mathcal{W}(c[N_i]) = w_i) \wedge (\mathcal{W}(c[N_j]) = w_j) \wedge (\mathcal{W}(c) = h)\}|. \tag{3.50}$$

The conditional CEP given that a codeword error results in all ith user's symbols received correctly while all jth user's symbols received erroneously is given by

$$\underline{\Phi_c}(\gamma) = \Phi_c \left(\sum_{h=d}^{n} O^{i,j}(0, n_j, h), \gamma \right) \tag{3.51}$$

where assuming that the all-zero codeword is transmitted we only considered codewords with a zero weight in the ith partition and a full weight in the jth partition.

In general, for a p-partition of the coordinates, let $\mathcal{P}$ and $\mathcal{Q}$ be the set of users (partitions) whose symbols are all received correctly and erroneously, respectively, in

---

[2]Conditional functions will have have the same notation as the unconditional ones except for an underbar.

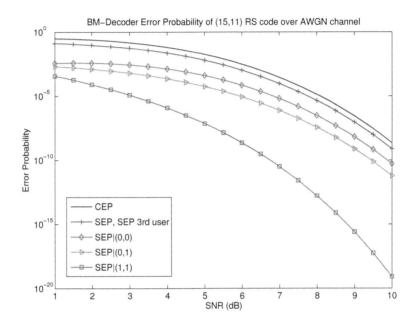

Figure 3.5: Conditional multiuser decoder error probability for Example 3.4.
For the Berlekamp-Massey decoder, the unconditional CEP and SEP are labeled
"CEP" and "SEP" respectively. The conditional SEP of the third user for cases 1,
2 and 3 are labeled "SEP$|(0,0)$," "SEP$|(0,1)$," and "SEP$|(1,1)$" respectively.

case of a codeword error. Let $\mathcal{O}$ be the set of users with no condition on their error
probability. The conditional error probability is calculated by considering only the
codewords which have a full weight for the coordinates in $\mathcal{Q}$ and a zero weight for
the coordinates in $\mathcal{P}$. By considering only such combinations in the sum of (3.2), the
conditional PWGF is derived as

$$
\mathbb{P}(\mathcal{X}_1, \mathcal{X}_2, \ldots, \mathcal{X}_p) = \sum_{i \in \Delta} \sum_{w_i=0}^{n_i} A(w_1, w_2, \ldots, w_p) \mathcal{X}_1^{w_1} \mathcal{X}_2^{w_2} \ldots \mathcal{X}_p^{w_p} \left| \begin{array}{ll} w_i = 0, & \text{if } i \in \mathcal{P} \\ w_i = n_i, & \text{if } i \in \mathcal{Q} \end{array} \right. .
$$

$$(3.52)$$

The conditional symbol error probability of the jth user is

$$\Phi_s^j(\gamma) = \Phi_c\left(Q^j(k,h), \gamma\right),\tag{3.53}$$

where $Q^j(k,h) = \sum_{w=1}^{n_j} \frac{w}{n_j} \underline{Q}^j(w,h)$ and $\underline{Q}^j(w,h)$ is the conditional IOWE of the jth partition and is derived from $\mathbb{P}(\mathcal{X}_1, \mathcal{X}_2, \ldots, \mathcal{X}_p)$ (see (3.7)). For example, if the first partition contains header information, then the conditional symbol error probability of the ith user given that the header is received correctly can be calculated by

$$\Phi_s^j(\gamma) = \Phi_c\left(\sum_{w=1}^{n_j} \frac{w}{n_j} O^{1,j}(0,w,h), \gamma\right).\tag{3.54}$$

Similarly, for bit-level decoding of the code's binary image, $\hat{Q}^j(mk,h)$ will be derived from $\tilde{\mathbb{P}}(\mathcal{X}_1, \mathcal{X}_2, \ldots, \mathcal{X}_p)$. If the users in $\mathcal{P}$ and $\mathcal{Q}$ have zero and one bit error probability respectively, the conditional binary PWGF only takes into account such codewords that have a zero binary weight for the partitions in $\mathcal{P}$ and a full binary Hamming weight for the partitions in $\mathcal{Q}$. The conditional BEP of the jth user follows by the substitution $\tilde{E}(h) := \hat{Q}^j(mk,h)$ in (3.42).

Example 3.4. Consider an systematic $(15, 11, 5)$ RS code and a partition $\mathcal{T} = (3, 3, 5, 4)$ of its coordinates where the last partition has the redundancy symbols and each of the first three partitions is assigned to a different user. The first partition may be assigned to be the header. Let the RS code be transmitted over an AWGN channel and decoded by a hard-decision bounded minimum distance (Berlekamp-Massey) decoder. From (3.33), (3.49) and Theorem 3.17 it follows that the unconditional CEP and SEP of any user is equal to the overall SEP and can be expressed as, respectively,

$$\Phi_c(\gamma) = \sum_{h=5}^{15} E(h) \sum_{t=0}^{\tau} P_t^h(\gamma),$$

$$\Phi_s(\gamma) = \sum_{h=5}^{15} \frac{h}{15} E(h) \sum_{t=0}^{\tau} P_t^h,$$

such that $E(h)$ is the weight enumerator as given by (3.10). The partition weight

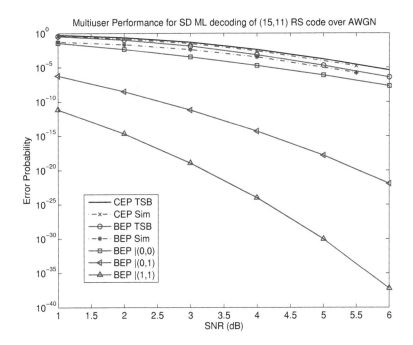

Figure 3.6: Conditional multiuser error probability for Example 3.5.
For the bit-level soft-decision maximum-likelihood decoder, the conditional bit error probability of cases 1, 2 and 3 are labeled "BEP$|(0,0)$," "BEP$|(0,1)$," and "BEP$|(1,1)$." The bounds on the unconditional CEP and BEP, labeled "CEP TSB" and "BEP TSB," are compared with the corresponding simulations, labeled "CEP Sim" and "BEP Sim," respectively.

generating function is given by

$$\mathbb{P}(\mathcal{W},\mathcal{X},\mathcal{Y},\mathcal{Z}) = \sum_{w_1=0}^{3} \sum_{w_2=0}^{3} \sum_{w_3=0}^{5} \sum_{w_4=0}^{4} A^T(w_1,w_2,w_3,w_4)\mathcal{W}^{w_1}\mathcal{X}^{w_2}\mathcal{Y}^{w_3}\mathcal{Z}^{w_4},$$

and the IOWGF of the third user is $\mathbb{O}^3(\mathcal{X},\mathcal{Y}) = \mathbb{P}(\mathcal{X},\mathcal{X},\mathcal{X}\mathcal{Y},\mathcal{X})$. We will now calculate the conditional symbol error probability of the third user under different scenarios.

Case 1:   The first two users have a zero error probability. Thus the PWGF conditioned on that the first two partitions have zero weight is

$$\underline{\mathbb{P}}_{(0,0)}(\mathcal{Y},\mathcal{Z}) = \sum_{w_3=0}^{5} \sum_{w_4=0}^{4} A^T(0,0,w_3,w_4)\mathcal{Y}^{w_3}\mathcal{Z}^{w_4}.$$

The conditional IOWGF of the third user is

$$\underline{\mathbb{O}}^3_{(0,0)}(\mathcal{X},\mathcal{Y}) = \underline{\mathbb{P}}_{(0,0)}(\mathcal{X}\mathcal{Y},\mathcal{Y}) = \sum_{w}\sum_{h} \underline{Q}^{1,2,3}(0,0,w,h)\mathcal{X}^w\mathcal{Y}^h,$$

It follows that the SEP of the third user conditioned on that the first two users have a zero error probability is

$$\underline{\Phi}^3_s(\gamma) = \sum_{h=d}^{n} \sum_{w=1}^{5} \frac{w}{5}\underline{Q}^{1,2,3}(0,0,w,j) \sum_{t=0}^{\tau} P^h_t.$$

Case 2:   The first and second users have an SEP of zero and one respectively. The corresponding conditional PWGF is

$$\underline{\mathbb{P}}_{(0,1)}(\mathcal{X},\mathcal{Y},\mathcal{Z}) = \sum_{w_3=0}^{5} \sum_{w_4=0}^{4} A^T(0,3,w_3,w_4)\mathcal{X}^3\mathcal{Y}^{w_3}\mathcal{Z}^{w_4}.$$

The corresponding IOWGF of the third user is

$$\underline{\mathbb{O}}^3_{(0,1)}(\mathcal{X},\mathcal{Y}) = \underline{\mathbb{P}}_{(0,1)}(\mathcal{Y},\mathcal{X}\mathcal{Y},\mathcal{Y}) = \sum_{w}\sum_{h} \underline{Q}^{1,2,3}(0,3,w,h)\mathcal{X}^w\mathcal{Y}^h.$$

To calculate the conditional SEP, we proceed as in the previous case.

Case 3: Both the first and second users have an SEP of one. The conditional SEP of the third user is

$$\underline{\Phi}_s^3(\gamma) = \sum_{h=d}^{n} \sum_{w=1}^{5} \frac{w}{5} \underline{Q}^{1,2,3}(3,3,w,j) \sum_{t=0}^{\tau} P_t^h.$$

where $\underline{Q}^{1,2,3}(3,3,w,h)$ is the coefficient of $\mathcal{X}^w\mathcal{Y}^h$ in $\underline{\mathbb{O}}_{(1,1)}^3(\mathcal{X},\mathcal{Y}) = \underline{\mathbb{P}}_{(1,1)}(\mathcal{Y},\mathcal{Y},\mathcal{X}\mathcal{Y},\mathcal{Y})$ and

$$\underline{\mathbb{P}}_{(1,1)}(\mathcal{W},\mathcal{X},\mathcal{Y},\mathcal{Z}) = \sum_{w_3=0}^{5} \sum_{w_4=0}^{4} A^{\mathcal{T}}(3,3,w_3,w_4)\mathcal{W}^3\mathcal{X}^3\mathcal{Y}^{w_3}\mathcal{Z}^{w_4}.$$

For an AWGN channel and a Berlekamp-Massey decoder, the codeword error probability, symbol error probability and the conditional symbol error probabilities for the third user for the three cases are plotted in Figure 3.5. It is observed that the conditional error probability of the third user given that other users have an error probability of one (Case 3) is the lowest compared to the other two cases. The reason is that in Case 3, one only considers errors due to the received word falling closer to codewords at a much larger Hamming distance from the transmitted one, and such an event happens with relatively lower probability.   ⬦

The same technique can be used to bound the performance of other symbol based decoders, such as the hard-decision maximum-likelihood decoder, under various scenarios. Next we consider analyzing the multiuser error probability when the decoder is a bit level decoder.

Example 3.5. Consider the $(15, 11, 5)$ code over $\mathbb{F}_{16}$ partitioned as in Example 3.4 and an SD bit-level ML decoder is employed at the output of an AWGN channel. The unconditional CEP and BEP are given by, respectively,

$$\Phi_c\left(\tilde{E}(h),\gamma\right) \leq \min_{\alpha}\left\{\sum_{h=5}^{\alpha} \tilde{E}(h)\mathcal{J}(\gamma,h) + \mathcal{G}(\gamma,\alpha)\right\},$$

$$\Phi_b(\gamma) = \min_{\alpha}\left\{\sum_{h=5}^{\alpha} \frac{h}{60}\tilde{E}(h)\mathcal{J}(\gamma,h) + \mathcal{G}(\gamma,\alpha)\right\},$$

where $\mathcal{J}(\gamma,h)$ and $\mathcal{G}(\gamma,\alpha)$ will be determined by the Poltyrev tangential sphere bound [87] (c.f., Section 8.1.3). We will now discuss the conditional bit error probability for different cases (as in Example 3.4):

Case 1: The first two users have a zero error probability. The average binary IOWE of the third user given the first two partitions have a zero weight is

$$\underline{\tilde{O}}^3_{(0,0)}(\mathcal{X}, \mathcal{Y}) = \underline{\tilde{\mathbb{P}}}_{(0,0)}(\mathcal{X}\mathcal{Y}, \mathcal{Y}) = \sum_{h=0}^{60} \sum_{w=0}^{20} \underline{\tilde{Q}}^{1,2,3}(0, 0, w, h)\mathcal{X}^w \mathcal{Y}^h,$$

such that $\underline{\tilde{\mathbb{P}}}_{(0,0)}(\mathcal{X}, \mathcal{Y}) = \underline{\mathbb{P}}_{(0,0)}(F(\mathcal{X}), F(\mathcal{Y}))$, and $F(\mathcal{X})$ is as defined in Theorem 3.14. The conditional BEP of the third user is given by

$$\underline{\Phi}^3_b(\gamma) = \min_\alpha \left\{ \sum_{h=5}^{\alpha} \sum_{w=1}^{20} \frac{w}{20} \underline{\tilde{Q}}^{1,2,3}(0, 0, w, h)\mathcal{J}(\gamma, h) + \mathcal{G}(\gamma, \alpha) \right\}.$$

Case 2: The first and second users have a zero and one bit error probability respectively. Let $\tilde{\mathbb{P}}(\mathcal{W}, \mathcal{X}, \mathcal{Y}, \mathcal{Z}) = \mathbb{P}(F(\mathcal{W}), F(\mathcal{X}), F(\mathcal{Y}), F(\mathcal{Z}))$ be the average binary PWGF then

$$\tilde{\mathbb{P}}_{(0,1)}(\mathcal{X}, \mathcal{Y}, \mathcal{Z}) = \text{Coeff}\left(\tilde{\mathbb{P}}(\mathcal{W}, \mathcal{X}, \mathcal{Y}, \mathcal{Z}), \mathcal{W}^0 \mathcal{X}^{12} \right) \mathcal{X}^{12},$$

and the conditional IOWE of the third user is

$$\tilde{Q}^{1,2,3}(0, 12, w, h) = \text{Coeff}\left(\tilde{\mathbb{P}}_{(0,1)}(\mathcal{Y}, \mathcal{X}\mathcal{Y}, \mathcal{Y}), \mathcal{X}^w \mathcal{Y}^h \right).$$

The conditional BEP is then given by

$$\underline{\Phi}^3_b(\gamma) = \min_\alpha \left\{ \sum_{h=5}^{\alpha} \sum_{w=1}^{20} \frac{w}{20} \underline{\tilde{Q}}^{1,2,3}(0, 12, w, h)\mathcal{J}(\gamma, h) + \mathcal{G}(\gamma, \alpha) \right\}.$$

Case 3: The average BEP of the first two users is one. In this case, the conditional PWGF can be calculated by

$$\tilde{\mathbb{P}}_{(1,1)}(\mathcal{W}, \mathcal{X}, \mathcal{Y}, \mathcal{Z}) = \text{Coeff}\left(\tilde{\mathbb{P}}(\mathcal{W}, \mathcal{X}, \mathcal{Y}, \mathcal{Z}), \mathcal{W}^{12}\mathcal{X}^{12} \right) \mathcal{W}^{12}\mathcal{X}^{12}.$$

One can then proceed to calculate the conditional IOWE and BPE of the third user by

$$\tilde{\underline{O}}^{1,2,3}(12,12,w,h) \;=\; \mathrm{Coeff}\left(\tilde{\mathbb{P}}_{(1,1)}(\mathcal{Y},\mathcal{Y},\mathcal{XY},\mathcal{Y}),\mathcal{X}^{w}\mathcal{Y}^{h}\right)$$

$$\underline{\Phi}_{b}^{3}(\gamma) = \min_{\alpha}\left\{\sum_{h=5}^{\alpha}\sum_{w=1}^{20}\frac{w}{20}\tilde{\underline{O}}^{1,2,3}(12,12,w,h)\mathcal{J}(\gamma,h) + \mathcal{G}(\gamma,\alpha)\right\}.$$

In Figure 3.6, the TSB on the codeword and bit error probability are plotted and compared to simulations of the ML decoder for a specific basis representation of the RS code. The conditional BEP of the third user is plotted for cases $1, 2$ and 3 . As in the previous example, it is observed that the conditional error probability of specific users given that some users have a high error probability decreases with the number of such users.   ◇

Example 3.6. Consider an systematic $(31,15,17)$ RS code over $\mathbb{F}_{32}$ and a partition $\mathcal{T} = (3,6,6,16)$ of its coordinates where the last partition has the redundancy symbols and each of the first three partitions is assigned to a different user. The first partition may be assigned to be the header. Let the binary image of a RS code be transmitted over an AWGN channel and decoded by a hard-decision symbol-based maximum-likelihood decoder decoder. We used the upper bound of Theorem 8.9 to bound the performance of the HD-ML decoder over $\mathbb{F}_{32}$. The CEP, SEP and conditional SEP are of the form of (3.32), (3.40) and (3.53). We consider three cases:

Case 1: The unconditional error probability of the third user.

Case 2: The symbol error probability of the third user given that the first user (header) is received correctly.

Case 3: The symbol error probability of the third user given that the first two users have their symbols received correctly.

The numerical results are shown in Figure 3.7. We observe that the unconditional CEP and SEP are very close. As more and more conditions are imposed, the conditional error probability of the third user decreases. *Case 2,* is of special interest, since in some cases the header will contain the routing information and it will be essential to estimate the error probability in case the information is routed correctly.   ◇

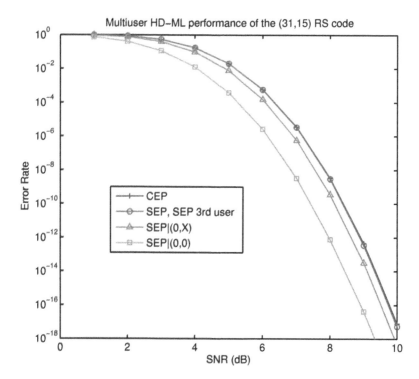

Figure 3.7: Conditional multiuser error probability of Example 3.6.
For the symbol-level hard-decision maximum-likelihood decoder of the $(31, 15)$ RS over $\mathbb{F}_{32}$, the unconditional CEP and SEP are plotted (Case 1). The conditional SEP of Case 2 and Case 3 are labeled "SEP$|(0, X)$" and "SEP$|(0, 0)$" respectively.

3.7   Conclusion

A closed form formula for the partition weight enumerator of maximum distance sep-arable (MDS) codes is derived. The average partition weight enumerator (PWE) is derived for the binary image of MDS codes defined over a field of characteristic two. We show that for MDS codes, all the coordinates have the same weight in the subcode composed of codewords with equal weight. We prove that a code has this property iff its dual code has this property. Consequently, it is shown that the first-order Reed-Muller codes and the extended Hamming codes have this property. A common approximation used to evaluate the symbol and bit error probabilities is thus shown to be exact for MDS codes. These results are employed to study the error probability when a Reed-Solomon code is used in a network scenario and is shared among different users. We show that MDS (e.g., RS) codes have many attractive features which makes their use in networks attractive. It is proved that the unconditional error probability of all the users will be the same regardless of the size of their partitions. As for the conditional error probabilities, they can be a useful measure in determining the performance of a user, if its performance depends on the correct transmission of a certain packet or header.

Chapter 4

Algebraic Soft-Decision Decoding of Reed-Solomon Codes: Interpolation Multiplicity Assignments

Simple things should be simple, complex things should be possible.

—Alan Kay

Reed-Solomon codes [93] are one of the most important types of error-correcting codes, due to their wide applicability in data-storage and communication systems. Through the seminal work of Sudan [102], Guruswami and Sudan [49], and Koetter and Vardy [72], we now have a polynomial-time algebraic soft-decision decoding (ASD) algorithm for Reed-Solomon codes. In an attempt to find asymptotic (in decoder complexity) performance limits for ASD, we develop a new class of *multiplicity assignment* algorithms for ASD in this chapter. Roughly speaking, the idea is to choose the multiplicity matrix so as to maximize the probability that the causal codeword is on the decoder's list, as suggested by [83], rather than to maximize the expected score of the causal codeword, as is done in [72]. However, whereas in [83], a Gaussian approximation is employed, we use a Chernoff bound instead. (It was independently suggested in [92], in a somewhat different context, to use the Chernoff bound in optimizing symbol based multiplicity matrices.)

Here is an overview of this chapter. Some preliminaries are given in Section 4.1. In Section 4.2, we give a brief overview of the Guruswami-Sudan (GS) algorithm. We also prove some interesting results that will become useful later in this chapter. In Section 4.4, we describe a mathematical framework for alebraic soft-decision decoding. A quick review of previously proposed multiplicity assignment algorithms for algebriac soft-decision decoding is given in Section 4.5. In Section 4.6, we formulate the multiplicity assignment problem as an optimization problem. Our algorithm is developed and explained in Section 4.7. We propose a Chernoff bound approach for the multiplicity assignment optimization problem. We study the cases of finite and infinite interpolation cost. We show that the formulated problem is convex and devise an iterative algorithm to solve it. In Section 4.8, we present some numerical results and discussions. We conclude the chapter and hint at future research directions in Section 4.10. Briefly, we conclude that our method is theoretically superior to previously proposed algebraic soft-decision algorithms, although whether it will prove to be practical remains to be seen.

4.1   Preliminaries

Throughout this chapter $\mathbb{F}_q$ will denote a finite field with q elements, and a typical element of $\mathbb{F}_q$ will be denoted by β. $\mathcal{C}$ will be an (n, k, d) Reed-Solomon code over $\mathbb{F}_q$.[1] Let the information data vector of k elements be $\boldsymbol{d} = (d_0, d_1, \ldots d_{k-1})$. Then the corresponding codeword $\boldsymbol{c} = (c_1, \ldots, c_n)$ is generated by polynomial evaluation of the information polynomial $\mathbb{D}(X) = \sum_{i=0}^{k-1} d_i X^i$ at n distinct nonzero elements of $\mathbb{F}_q$ constituting the support set of the code, $S = \{s_i; s_i \in \mathbb{F}_q \text{ for } i = 1, 2, \ldots, n\}$. That is $c_i = \mathbb{D}(s_i)$ for $i = 1, 2, \ldots, n$.

We will often encounter $q \times n$ arrays (or matrices) of real numbers, typically denoted by $W = (w_i(\beta))$, where $i = 1, \ldots, n$ and $\beta \in \mathbb{F}_q$. The cost of such an array is defined to be

$$\Omega(W) \triangleq \frac{1}{2} \sum_{i=1}^{n} \sum_{\beta \in \mathbb{F}_q} w_i(\beta) \left(w_i(\beta) + 1 \right). \tag{4.1}$$

---

[1] More precisely, $\mathcal{C}$ may be a coset of the parent RS code. See Corollary 4.4.

If $\boldsymbol{u} = (u_1, \ldots, u_n)$ is a n-dimensional vector over $\mathbb{F}_q$, the *score* of $\boldsymbol{u}$ with respect to the array W is is defined to be

$$\langle \boldsymbol{u}, W \rangle \triangleq \sum_{i=1}^{n} w_i(u_i). \tag{4.2}$$

The underlying (discrete input, memoryless) channel model has input alphabet $\mathbb{F}_q$, output alphabet R (which could be of infinite size for continuous channels), and transition probabilities $\Pr\{Y = r | X = \beta\}$, where X and Y denote the channel input and output respectively. Given a received symbol $r \in R$, there is a unique *a posteriori* density function on $\mathbb{F}_q$ corresponding to each $\beta \in \mathbb{F}_q$;

$$p_r(\beta) = \Pr\{X = \beta | Y = r\}.$$

Observing a channel output r is therefore equivalent to being given $p_r(\beta)$ for all $\beta \in \mathbb{F}_q$. From this viewpoint, the output alphabet is not R but

$$\mathcal{R} = \{p_r(\beta); r \in R, \beta \in \mathbb{F}_q\}.$$

Thus in this chapter we will assume that if $\boldsymbol{c} = (c_1, \ldots, c_n)$ is transmitted, the received word is an array of density functions $\Pi = (\pi_i(\beta))$, where $\pi_i(\beta) \in \mathcal{R}$, for $i = 1, \ldots, n$ and $\beta \in \mathbb{F}_q$. We call Π the *a posteriori probability*, or APP, matrix. We denote by $\overline{\mathcal{R}}$ the set of all possible APP matrices. It should be noted that the density functions $\pi_i(\beta)$ could be calculated from the soft channel output as is the case for additive white Gaussian noise (AWGN) channels. However, the density functions could also be delivered directly as the soft output of an inner decoder such as the BCJR algorithm [7] or the soft output Viterbi algorithm (SOVA) [51, 112] in concatenated coding systems.

The indicator function Δ is defined to be

$$\Delta\,[\texttt{condition}] = \begin{cases} 1, & \text{if } \texttt{condition is true} \\ 0, & \text{if } \texttt{condition is false} \end{cases}. \tag{4.3}$$

We will denote the ubiquitous quantity $(k-1)$ by v. We will finish this section by giving some definitions that are crucial to understanding the GS algorithm [76].

Definition 4.1. The (r, s)th Hasse derivative of a bivariate polynomial $\mathbb{B}(X, Y) =$

$\sum_{i,j} B_{i,j} X^i Y^j$ at (α, β) is given by

$$
\begin{aligned}
B'_{r,s}(\alpha, \beta) &= \mathrm{Coeff}(\mathbb{B}(X + \alpha, Y + \beta), X^r Y^s) \\
&= \sum_{i,j} \binom{i}{r}\binom{j}{s} B_{i,j} \alpha^{i-r} \beta^{j-s},
\end{aligned}
$$

where the coefficient function is defined by $B_{i,j} = \mathrm{Coeff}(\mathbb{B}(X,Y), X^i Y^j)$.

Definition 4.2. The bivariate polynomial $\mathbb{B}(X,Y)$ passes through the point (α, β) with multiplicity m (has a zero of multiplicity m at (α, β)) iff

$$
B'_{r,s}(\alpha, \beta) = 0 \text{ for all } r \text{ and } s \text{ such that } 0 \le r + s < m,
$$

equivalently, iff $\mathbb{B}(X + \alpha, Y + \beta)$ does not contain any monomial of degree strictly less than m.

Definition 4.3. The (w_x, w_y)-weighted degree of a bivariate polynomial $\mathbb{B}(X,Y) = \sum_{i,j} B_{i,j} X^i Y^j$

$$
\deg_{w_x, w_y} \mathbb{B}(X,Y) \triangleq \max\{iw_x + jw_y \; : \; B_{i,j} \ne 0\}.
$$

It follows that X-degree $\deg_X \mathbb{B}(X,Y) = \deg_{1,0} B(X,Y)$, the Y-degree $\deg_Y \mathbb{B}(X,Y) = \deg_{0,1} \mathbb{B}(X,Y)$ and the total degree $\deg \mathbb{B}(X,Y) = \deg_{1,1} \mathbb{B}(X,Y)$.

4.2   The Guruswami-Sudan Algorithm

Given a $q \times n$ array of nonnegative integers $M = (m_i(\beta))$, called a *multiplicity matrix*, associated with an $(n, v + 1, d)$ Reed-Solomon code, the (modified) GS algorithm is a list-decoding algorithm consisting of two main steps [49, 76]

1. *Interpolation:* Construct a bivariate polynomial, $\mathbb{B}(X,Y)$, of minimum $(1, v)$-weighted degree that passes through each of the points (s_i, β) with multiplicity $m_i(\beta)$, where $\beta \in \mathbb{F}_q$ and $i = 1, 2, \ldots, n$.

2. *Factorization:* Find all linear factors of $\mathbb{B}(X,Y)$, $(Y - \mathbb{G}(X))|\mathbb{B}(X,Y)$, where $\mathbb{G}(X)$ is a polynomial of degree less than or equal to v. The codeword corresponding to each such polynomial $\mathbb{G}(X)$ is placed on the list.

The GS algorithm produces as an output a list of at most $\sqrt{2\,\Omega(M)/v}$ codewords [76], which contains all codewords $\boldsymbol{c}$ such that

$$\langle \boldsymbol{c}, M \rangle > D_v(\Omega(M)), \tag{4.4}$$

where $D_v(\gamma)$ is the least positive integer D such that

$$\left| \left\{ (i,j) \in \mathbb{N}^2; i + vj \le D \right\} \right| \ge \gamma + 1.$$

In other words, $D_v(\Omega(M))$ is the minimal $(1,v)$-weighted degree of a bivariate polynomial $\mathbb{B}(X,Y)$ in order for such a nontrivial polynomial, that could be interpolated to pass through all the points (s_i, β) with multiplicity at least $m_i(\beta)$, exists. If the sufficient condition of (4.4) is satisfied, then the bivariate polynomial $\mathbb{B}(X,Y)$ will have a linear factor of the form $Y - \mathbb{G}(X)$ where $\mathbb{G}(X)$ has a degree at most v and is the data polynomial associated with the codeword $\boldsymbol{c}$ [49, 72].

One can show that for an interpolation cost γ, the minimum $(1,v)$-weighted degree admits to this closed form formula

$$D_v(\gamma) = \left\lfloor \frac{\gamma}{m} + \frac{v(m-1)}{2} \right\rfloor, \quad \text{where } m = \left\lfloor \sqrt{\frac{2\gamma}{v} + \frac{1}{4}} + \frac{1}{2} \right\rfloor. \tag{4.5}$$

If complexity is not issue and the interpolation cost tends to infinity, then a sufficient condition of (4.4) for a codeword $\boldsymbol{c}$ to be on the GS list reduces to [72, 31] (see Theorem 4.7)

$$\frac{\langle \boldsymbol{c}, M \rangle}{\|M\|_2} > \sqrt{v}. \tag{4.6}$$

In the rest of this chapter, we will denote the sufficient condition of (4.4) by

$$\boldsymbol{c} \vdash M. \tag{4.7}$$

4.3 Upper Bounds on the Minimum Weighted Degree

In this section, we give some technical results needed later. Whereas the discrete function $D_v(\gamma)$ can be calculated by the closed form formula of (4.5), it will be more convenient if we can have continuous tight upper bounds on $D_v(\gamma)$.

Lemma 4.1. *An upper bound on the function $D_v(\gamma)$ is*

$$D_v(\gamma) \leq -\frac{v}{2} + \sqrt{2v\gamma} + \frac{v^{3/2}}{8\sqrt{2\gamma}}. \tag{4.8}$$

Proof. Let m be the unique integer satisfying [80]

$$\binom{m}{2} \leq \frac{\gamma}{v} < \binom{m+1}{2}. \tag{4.9}$$

Thus, $\gamma \geq \frac{vm(m-1)}{2}$. Let $\psi(m) = \frac{\gamma}{m} + \frac{v(m-1)}{2}$, then $\psi(m) \geq v(m-1)$. Thus,

$$\frac{\partial \psi(m)}{\partial m} \geq v \geq 0,$$

which implies that $\psi(m)$ is a nondecreasing function of m if γ satisfies (4.9). Since $m \leq \left(\sqrt{\frac{2\gamma}{v} + \frac{1}{4}} + \frac{1}{2}\right)$, it follows that

$$D_v(\gamma) = \lfloor \psi(m) \rfloor \leq \psi(m) \leq \psi\left(\sqrt{\frac{2\gamma}{v} + \frac{1}{4}} + \frac{1}{2}\right). \tag{4.10}$$

With some algebra, we get

$$D_v(\gamma) \leq -\frac{v}{2} + \sqrt{2v\gamma + \frac{v^2}{4}} \leq -\frac{v}{2} + \sqrt{2v\gamma}\left(1 + \frac{v}{16\gamma}\right), \tag{4.11}$$

which implies the assertion. □

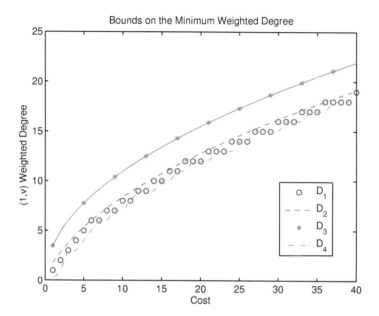

Figure 4.1: Bounds on the function $D_v(\Omega(M))$ as a function of $\Omega(M)$ for $v = 6$. The bounds D_1, D_2, D_3, D_4 are given by equations (4.5), (4.8), (4.12), (4.13) respectively.

From the derivation of the above lemma it is clear that the upper bound of [72]

$$D_v(\gamma) \leq \sqrt{2v\gamma} \qquad (4.12)$$

is a looser upper bound than that of (4.8). In fact, the function $D_v(\gamma)$ is well approximated by

$$D_v(\gamma) \approx \left\lfloor \sqrt{2v\gamma} - \frac{v}{2} \right\rfloor. \qquad (4.13)$$

Indeed, if v is fixed, $0 \leq D_v(\gamma) - \lfloor \sqrt{2v\gamma} - \frac{v}{2} \rfloor \leq 1$ for all sufficiently large γ. In Figure 4.1, the discrete function $D_v(\Omega(M))$ is plotted for $v = 6$ as a function of the cost $\Omega(M)$. The upper bounds of (4.8) and (4.12) are also plotted. It is clear that the upper bound of (4.8) is a tight (continuous) upper bound. The approximation of (4.13) is also compared to the function $D_v(\Omega(M))$.

Lemma 4.2. *If $\gamma > 0$,*
$$\lim_{\lambda \to \infty} \frac{D_v(\lambda^2 \gamma)}{\lambda} = \sqrt{2v\gamma}.$$

Proof. Using (4.8), $\lim_{\lambda \to \infty} \frac{D_v(\lambda^2 \gamma)}{\lambda} = \lim_{\lambda \to \infty} \frac{-v}{2\lambda} + \frac{\lambda\sqrt{2v\gamma}}{\lambda} = \sqrt{2v\gamma}.$ □

4.4 A Mathematical Model for ASD Decoding of Reed-Solomon Codes.

In this section we describe a model for algebraic soft–decision decoding of RS codes. A codeword $c = (c_1, \ldots, c_n)$ which we call the *causal codeword*, is selected at random from $\mathcal{C}$, transmitted over a memoryless channel, and received as the APP matrix $\Pi = (\pi_i(\beta))$ where $i = 1, \ldots, n$ and $\beta \in \mathbb{F}_q$. Given the APP matrix Π, the ASD decoding algorithm converts Π into a $q \times n$ multiplicity matrix M. This multiplicity matrix is forwarded to the GS algorithm, which in turn produces a list of codewords. If $c \vdash M$, then the causal codeword c will be on the list in which case the decoder is declared to have succeeded.

The situation is summarized by the following chain of random vectors and matrices:[2]

$$c \to \Pi \xrightarrow{A} M. \tag{4.14}$$

The only quantity in (4.14) under engineering control is the multiplicity algorithm A, so the problem of optimizing the ASD algorithm is equivalent to choosing the right multiplicity algorithm:

$$P(\mathcal{A}) = \min_{A \in \mathcal{A}} \Pr \{\mathcal{E}_A\}, \tag{4.15}$$

where

$$\mathcal{E}_A = \{c \nvdash M\}, \tag{4.16}$$

---

[2]In order to minimize our notational complexity, we do not distinguish notationally between a random variable and an instance of the random variable.

and $\mathcal{A}$ is some suitably restricted class of multiplicity algorithms. Note that

$$\Pr\{\mathcal{E}_A\} = \sum_{\Pi \in \overline{\mathcal{R}}} \Pr\{\mathcal{E}_A | \Pi\} \Pr\{\Pi\}, \tag{4.17}$$

so that A minimizes $\Pr\{\mathcal{E}_A\}$ iff it minimizes $\Pr\{\mathcal{E}_A|\Pi\}$ for each APP matrix Π. The following theorem shows that $\Pr\{\mathcal{E}_A|\Pi\}$ depends only on $\mathcal{C}$, Π and M, and so we introduce the notation

$$P_{\mathcal{C}}(\Pi, M) \stackrel{\Delta}{=} \Pr\{\mathcal{E}_A|\Pi\}.$$

Theorem 4.3. *For* $\boldsymbol{x} = (x_1, \ldots, x_n) \in \mathbb{F}_q^n$ *define* $\boldsymbol{P}(\boldsymbol{x}) = \prod_{i=1}^n \pi_i(x_i)$ *and* $\boldsymbol{P}(\mathcal{C}) = \sum_{\boldsymbol{c} \in \mathcal{C}} \boldsymbol{P}(\boldsymbol{c})$. *Then*

$$P_{\mathcal{C}}(\Pi, M) = \frac{1}{\boldsymbol{P}(\mathcal{C})} \sum_{\boldsymbol{c} \in \mathcal{C}} \Delta\left[\boldsymbol{c} \nvdash M\right] \boldsymbol{P}(\boldsymbol{c}). \tag{4.18}$$

Proof. First,

$$\Pr\{\mathcal{E}_A|\Pi\} = \sum_{\boldsymbol{c} \in \mathcal{C}} \Delta\left[\boldsymbol{c} \nvdash M\right] \Pr\{\boldsymbol{c}|\Pi\}.$$

Second (c.f., [72], Appendix A)

$$\Pr\{\boldsymbol{c}|\Pi\} = \frac{\boldsymbol{P}(\boldsymbol{c})}{\boldsymbol{P}(\mathcal{C})}.$$

$\square$

In Theorem 4.3, it was implicitly assumed that the channel is memoryless and that the components of $\boldsymbol{c}$ are uniformly drawn from the field $\mathbb{F}_q$. But because of the maximal distance separable (MDS) property of RS codes, the elements of any subset of k or fewer components of $\boldsymbol{c}$ are independent and could be treated as information symbols. However, minimizing $P_{\mathcal{C}}(\Pi, M)$ directly is not easy due to the difficulty of calculating $\boldsymbol{P}(\mathcal{C})$ for an arbitrary code $\mathcal{C}$ and an arbitrary reliability matrix Π. But the following trick, due essentially to Koetter and Vardy [72], allows us to replace the Markov chain (4.14) with

$$\boldsymbol{x} \to \Pi \xrightarrow{A} M, \tag{4.19}$$

which is identical to (4.14) except that the random codeword drawn uniformly from the code $\boldsymbol{c} \sim U[\mathcal{C}]$ in (4.14) has been replaced with a random vector $\boldsymbol{x} \sim U[\mathbb{F}_q^n]$ in

(4.19), whose components are independent, where $\boldsymbol{x} \sim U[\mathcal{X}]$ means that $\boldsymbol{x}$ is drawn uniformly at random from the space $\mathcal{X}$.

Corollary 4.4. *If* $\mathcal{C}_1, \ldots, \mathcal{C}_K$ *are the cosets of* $\mathcal{C}$, *with* $K = q^{n-k}$, *then*

$$\sum_{i=1}^{K} \boldsymbol{P}(\mathcal{C}_i) P_{\mathcal{C}_i}(\Pi, M) = \sum_{\boldsymbol{x} \in \mathbb{F}_q^n} \Delta\left[\boldsymbol{x} \nvdash M\right] \boldsymbol{P}(\boldsymbol{x}) \qquad (4.20)$$

$$\triangleq \mathcal{P}(\Pi, M).$$

Since the left-hand side is an average of the error probability $P_{\mathcal{C}_i}(\Pi, M)$, *then* $P_{\mathcal{C}_i}(\Pi, M) \leq \mathcal{P}(\Pi, M)$ *for at least one coset* $\mathcal{C}_i$.

4.5   Algebraic Soft-Decision Decoding

As mentioned in the previous sections, the Guruswami-Sudan algorithm will take as an input a *multiplicity* matrix, $M = (m_i(\beta))$ and will output a list of codewords. Algorithms for assigning interpolation multiplicities for the GS algorithm were proposed based on different criteria [49, 72, 83, 85]. Before proceeding to derive our multiplicity assignment algorithm, we will briefly review two algorithms of particular interest.

The Koetter-Vardy Algorithm: The Koetter-Vardy algorithm finds the multiplicity matrix M that maximizes the expectation of the score, $E\{\langle \boldsymbol{x}, M \rangle\}$, where $\boldsymbol{x} \sim U[\mathbb{F}_q^n]$ is an n-dimensional random vector of independent components [72]. A reduced complexity KV algorithm is [46]

$$m_i(\beta) = \lfloor \lambda \pi_i(\beta) \rfloor, \qquad (4.21)$$

where $\lambda > 0$ is a complexity parameter determined by $\Omega(M)$. For $\Omega(M) = \gamma$, it can be shown that $\lambda = (-1 + \sqrt{1 + 8\gamma/n})/2$. In case of infinite interpolation cost, the sufficient condition of (4.6) reduces to

$$\frac{\langle \boldsymbol{c}, \Pi \rangle}{\|\Pi\|_2} > \sqrt{v}. \qquad (4.22)$$

The Gaussian Approximation: By the definition of the score, (4.2), the score of a random vector with respect to a multiplicity matrix M is a sum of n random variables.

Assuming that the n random variables are independent, the distribution of the score is approximated by a Gaussian distribution. Based on this approximation, an iterative algorithm is derived to find the multiplicity matrix of infinite interpolation cost that will minimize the error probability [83]. Note however that this approximation is valid if n is sufficiently large. The Gaussian approximation has been derived assuming infinite interpolation costs and it is not clear how to extend it to practical finite interpolation costs. The Gaussian approximation is also discussed in Section 4.9.

4.6 Optimum Multiplicity Matrices

In view of Corollary 4.4, in the rest of the chapter we will focus on choosing M so as to minimize $\mathcal{P}(\Pi, M)$, with the understanding that upper bounds on $\mathcal{P}(\Pi, M)$ technically apply only to the best cosets of the parent RS code.

4.6.1 Optimization Problem

Usually the ASD decoder will have a cost restriction, so we introduce the notation

$$P(\Pi, \gamma) = \min_{\Omega(M) \leq \gamma} \mathcal{P}(\Pi, M) \tag{4.23}$$

$$M(\Pi, \gamma) = \arg_M \min_{\Omega(M) \leq \gamma} \mathcal{P}(\Pi, M). \tag{4.24}$$

Here $P(\Pi, \gamma)$ is the minimum possible ASD decoder error probability, given Π and an upper bound of γ on the cost of M. The matrix $M(\Pi, \gamma)$ is the optimal multiplicity matrix of cost less than or equal to γ corresponding to the APP matrix Π.

We also define

$$P(\Pi, \infty) \triangleq \lim_{\gamma \to \infty} P(\Pi, \gamma), \tag{4.25}$$

which is the minimum possible decoder error probability, given the APP matrix Π, without regard to cost.

Finally, let us consider (c.f., (4.15)) the problem of computing

$$P(\gamma) \triangleq \min_{\Omega(M) \leq \gamma} \Pr\{\mathcal{E}_A\}, \tag{4.26}$$

the minimum possible ASD decoder error probability for decoder cost $\leq \gamma$, and

$$P(\infty) \triangleq \lim_{\gamma \to \infty} P(\gamma), \tag{4.27}$$

the absolute minimum ASD decoder error probability, regardless of cost. By (4.17) we have

$$P(\gamma) = \sum_{\Pi \in \mathcal{R}} P(\Pi, \gamma) \Pr\{\Pi\} \tag{4.28}$$

$$P(\infty) = \sum_{\Pi \in \mathcal{R}} P(\Pi, \infty) \Pr\{\Pi\}. \tag{4.29}$$

4.6.2   Soft Multiplicity Matrices: A Relaxation

It is difficult to deal with the requirement that the entries of M are integers, so we now define a slightly different problem in which the integer constraint is relaxed and the multiplicities can be arbitrary (nonnegative) real numbers.

Thus let $Q = (q_i(\beta))$ be a "soft" multiplicity matrix, i.e., for each $i = 1, \dots, n$, and each $\beta \in \mathbb{F}_q$, $q_i(\beta)$ is a nonnegative real number. We define

$$\mathcal{P}(\Pi, Q) \triangleq \sum_{\boldsymbol{x} \in \mathbb{F}_q^n} \Delta\left[\boldsymbol{x} \nvdash Q\right] \boldsymbol{P}(\boldsymbol{x}) \tag{4.30}$$

$$P^*(\Pi, \gamma) \triangleq \min_{\Omega(Q) \leq \gamma} \mathcal{P}(\Pi, Q) \tag{4.31}$$

$$Q^*(\Pi, \gamma) \triangleq \arg\min_{\Omega(Q) \leq \gamma} \mathcal{P}(\Pi, Q) \tag{4.32}$$

$$P^*(\Pi, \infty) \triangleq \lim_{\gamma \to \infty} P^*(\Pi, \gamma). \tag{4.33}$$

These quantities are the same as the corresponding unstarred ones, (4.23), (4.24), and (4.25), except that the integral matrices (with integer elements) M are replaced with real matrices Q, so that logically

$$P^*(\Pi, \gamma) \leq P(\Pi, \gamma) \tag{4.34}$$

$$P^*(\Pi, \infty) \leq P(\Pi, \infty). \tag{4.35}$$

Surprisingly, if cost is no object, we loose nothing by relaxing the constraint that the multiplicities be integers. In the following lemma, we show that up-scaling a multiplicity matrix Q with a scalar $\lambda > 1$, results in a lower error probability at the expense of a larger interpolation cost.

Lemma 4.5. *For any* (Π, Q),

$$\lim_{\lambda \to \infty} \mathcal{P}(\Pi, \lambda Q) \leq \mathcal{P}(\Pi, Q). \tag{4.36}$$

Proof. Suppose $\Delta\left[\boldsymbol{x} \vdash Q\right] = 1$, then with high probability this implies that

$$\langle \boldsymbol{x}, Q \rangle \geq \sqrt{2v\,\Omega(Q)} \tag{4.37}$$

for reasonably large costs $\Omega(Q)$. If $\lambda \geq 1$, $|\lambda Q| \leq \lambda^2 \Omega(Q)$, and

$$\frac{\langle \boldsymbol{x}, \lambda Q \rangle}{D_v(|\lambda Q|)} \geq \frac{\lambda \langle \boldsymbol{x}, Q \rangle}{D_v(\lambda^2 \Omega(Q))}. \tag{4.38}$$

But by Lemma 4.2, the limit of the right-hand side of (4.38) is $\langle \boldsymbol{x}, Q \rangle / \sqrt{2v\Omega(Q)} \geq 1$, with high probability, where the last inequality follows from (4.37). Thus

$$\lim_{\lambda \to \infty} \Delta\left[\boldsymbol{x} \vdash \lambda Q\right] = 1.$$

It follows that for any $\boldsymbol{x}$,

$$\lim_{\lambda \to \infty} \left\{ \sum_{\boldsymbol{x} \in \mathbb{F}_q^n} \Delta\left[\boldsymbol{x} \nvdash \lambda Q\right] \boldsymbol{P}(\boldsymbol{x}) \right\} \leq \sum_{\boldsymbol{x} \in \mathbb{F}_q^n} \Delta\left[\boldsymbol{x} \nvdash Q\right] \boldsymbol{P}(\boldsymbol{x}). \tag{4.39}$$

Comparing this to (4.30), we are done.   □

Theorem 4.6. $P^*(\Pi, \infty) = P(\Pi, \infty)$

Proof. Define P^+ to denote rational matrices. Then

$$P^*(\Pi, \infty) = P^+(\Pi, \infty), \tag{4.40}$$

by continuity, and

$$P^+(\Pi, \infty) = P(\Pi, \infty), \tag{4.41}$$

by the following argument. If Q is rational, then λQ is integral for arbitrarily large values of λ. Then Lemma 4.5 and (4.35) imply (4.41). $\qquad \square$

4.7 The Chernoff Bound Multiplicity Assignment Algorithm

In this section, we devise an interpolation assignment algorithm based on minimizing a tight upper bound on the error probability, the Chernoff bound.

4.7.1 The Chernoff Bound—Finite Cost

We have seen that the number $P^*(\Pi, \gamma)$ (see (4.31), above), delimits the best possible ASD decoding performance, if the APP matrix Π is given. Unfortunately, however, it is very difficult to compute $P^*(\Pi, \gamma)$. In this section, we derive a Chernoff bound on $P^*(\Pi, \gamma)$ (see (4.50), below), which is easy to compute.

Let $(\mathbb{F}_q^n, \Pi)$ be a discrete sample space, i.e., for $\boldsymbol{x} = (x_1, \ldots, x_n) \in \mathbb{F}_q^n$ and $\Pi = (\pi_i(\beta))$ define the probability measure $\boldsymbol{P}(\boldsymbol{x}) = \prod_{i=1}^{n} \pi_i(x_i)$. Define (independent) random variables $\mathcal{S}_1, \ldots, \mathcal{S}_n$ by

$$\mathcal{S}_i(\boldsymbol{x}) = q_i(x_i) \quad \text{for } i = 1, \ldots, n. \tag{4.42}$$

where $Q = ((q_i(\beta)))$ is the multiplicity matrix, and the score

$$S_Q = \langle \boldsymbol{x}, Q \rangle = \mathcal{S}_1 + \cdots + \mathcal{S}_n. \tag{4.43}$$

Now we have

$$\Pr\{S_Q \leq \delta\} = \sum_{\boldsymbol{x} \in \mathbb{F}_q^n} \Delta\left[\langle \boldsymbol{x}, Q \rangle \leq \delta\right] \boldsymbol{P}(\boldsymbol{x}). \tag{4.44}$$

Let $\phi_i(s, \pi_i, q_i)$ be the moment generating function for $\mathcal{S}_i$, i.e.,

$$\phi_i(s, \pi_i, q_i) = E_{\mathcal{S}_i}\left\{e^{s\mathcal{S}_i}\right\} = \sum_{\beta \in \mathbb{F}_q} \pi_i(\beta) e^{sq_i(\beta)}. \tag{4.45}$$

Then the moment generating function for S_Q is

$$\Phi(s, \Pi, Q) = \sum_t \Pr\{S_Q = t\} e^{st} = E_{S_Q}\left\{e^{sS_Q}\right\} \tag{4.46}$$

$$= E_{S_Q}\left\{e^{s\sum_{i=1}^n \mathcal{S}_i}\right\} = E_{S_Q}\left\{\prod_{i=1}^n e^{s\mathcal{S}_i}\right\} \tag{4.47}$$

$$= \prod_{i=1}^n E_{\mathcal{S}_i}\left\{e^{s\mathcal{S}_i}\right\} = \prod_{i=1}^n \phi_i(s, \pi_i, q_i), \tag{4.48}$$

where the expectation and the product are interchanged due to the assumption that the random variables $\mathcal{S}_i$ are independent. Then by the the Chernoff bound (c.f., [89, 118]),

$$\Pr\{S_Q \leq \delta\} = \sum_{t \leq \delta} \Pr\{S_Q = t\} \tag{4.49}$$

$$\leq \min_{s \geq 0}\left\{\sum_t \Pr\{S_Q = t\} e^{s(\delta-t)}\right\} = \min_{s \geq 0}\left\{e^{s\delta}\Phi(-s, \Pi, Q)\right\}.$$

Finally, if we recall that $P^*(\Pi, \gamma) \triangleq \min_{\Omega(Q) \leq \gamma} \mathcal{P}(\Pi, Q)$ we have

$$P^*(\Pi, \gamma) \leq P^X(\Pi, \gamma) \triangleq \min_{\substack{s \geq 0 \\ \Omega(\bar{Q})=\gamma}} \left\{e^{sD_v(\gamma)}\Phi(-s, \Pi, Q)\right\}. \tag{4.50}$$

It is a bit awkward to deal with the constraint $\Omega(Q) = \gamma$ in (4.50). We could replace this constraint with the more natural constraint $\|\mathbb{X}\|^2 = \sum_{i,\beta} X_i(\beta)^2 = L^2$, where $\mathbb{X} = (X_i(\beta))$ is of the same size as Q, by the following transformation:

$$X_i(\beta) = q_i(\beta) + 1/2; \quad L^2 = 2\gamma + \frac{nq}{4}; \quad D' = D_v(\gamma) + \frac{n}{2}. \tag{4.51}$$

Thus (4.50) could be written as

$$P^*(\Pi, \gamma) \leq \min_{\|\mathbb{X}\|^2 = L^2} \min_{s \geq 0} \left\{ e^{sD'} \Phi(-s, \Pi, \mathbb{X}) \right\},$$ (4.52)

and the optimum matrix is given by

$$\mathbb{X}^* = \arg_{\mathbb{X}} \min_{\|\mathbb{X}\|^2 = L^2} \min_{s \geq 0} \left\{ e^{sD'} \Phi(-s, \Pi, \mathbb{X}) \right\}.$$ (4.53)

4.7.2 The Chernoff Bound—Infinite Cost

In this section, we derive a methodology for performance analysis at asymptotically large costs. We begin by defining an auxiliary function $G^*(\Pi, \zeta)$:

$$G^*(\Pi, \zeta) = \min_{\|R\|^2 = 1} \sum_{\boldsymbol{x} \in \mathbb{F}_q^n} \Delta\left[\langle \boldsymbol{x}, R \rangle \leq \zeta \right] P(\boldsymbol{x}).$$ (4.54)

In the following theorem, we shall see that the case of $\gamma \to \infty$ is the special case of $L^2 = 1$ and $D' = \sqrt{v}$.

Theorem 4.7. $P^*(\Pi, \infty) = \lim_{\gamma \to \infty} P^*(\Pi, \gamma) = G^*(\Pi, \sqrt{v})$.

Proof. Define $R = \mathbb{X}/\|\mathbb{X}\|$, then $\|R\|^2 = 1$. By using (4.51) and Lemma 4.1,

$$\lim_{\gamma \to \infty} \frac{D'}{L} = \lim_{\gamma \to \infty} \frac{\sqrt{v} + \frac{v^{3/2}}{16\gamma} + \frac{n-v}{2\sqrt{2\gamma}}}{\sqrt{1 + \frac{nq}{8\gamma}}}.$$ (4.55)

Specifically, for large γ the right-hand side of (4.55) is approximated by

$$\sqrt{v} + \frac{v^{3/2}}{16\gamma} + \frac{n-v}{2\sqrt{2\gamma}} + \left(\sqrt{v} + \frac{v^{3/2}}{16\gamma} + \frac{n-v}{2\sqrt{2\gamma}} \right) \left(-\frac{1}{2}\frac{nq}{8\gamma} + \frac{1.3}{2.4}\left(\frac{nq}{8\gamma}\right)^2 + \ldots \right)$$
$$\to \sqrt{v} + o(1),$$

where $o(1) \to 0$ as $\gamma \to \infty$. Thus,

$$\lim_{\gamma \to \infty} \min_{\|\mathbb{X}\|^2 = L^2} \Pr\left\{ S_{\mathbb{X}} \leq D' \right\} = \lim_{\gamma \to \infty} \min_{\|R\|^2 = 1} \Pr\left\{ S_R \leq D'/L \right\} = \min_{\|R\|^2 = 1} \Pr\left\{ S_R \leq \sqrt{v} \right\}$$

which by comparing with (4.52) implies the assertion. □

Corollary 4.8. $P(\Pi, \infty) = P^*(\Pi, \infty) = G^*(\Pi, \sqrt{v})$.

Proof. By Theorem 4.6 and Theorem 4.7 we are done. □

Thus $G^*(\Pi, \sqrt{k-1})$ is the minimum possible decoder error probability for the ASD decoder, given the APP matrix Π. Similarly,

$$P(\infty) = \sum_{\Pi \in \mathcal{R}} G^*(\Pi, \sqrt{k-1}) \Pr\{\Pi\}, \tag{4.56}$$

is the unconditional minimum possible decoder error probability. The quantity $G^*(\Pi, \sqrt{v})$, like its finite-cost counterpart $P^*(\Pi, \gamma)$, is difficult to compute exactly, but easy to approximate with the Chernoff bound. To summarize: suppose $R = (r_i(\beta))$, with $\|R\|^2 = 1$ is given. On the $\{\mathbb{F}_q^n, \Pi\}$ sample space, define corresponding random variables $\mathcal{R}_i = r_i(x_i)$, for $i = 1, \ldots, n$. Then

$$G^*(\Pi, \zeta) = \min_{\|R\|^2=1} \Pr\{\mathcal{R}_1 + \cdots + \mathcal{R}_n \leq \zeta\}. \tag{4.57}$$

Let

$$\gamma_i(s, \pi_i, r_i) = \sum_{x \in \mathbb{F}_q} \pi_i(x) e^{s r_i(x)} \tag{4.58}$$

be the moment generating function for $\mathcal{R}_i$, $i = 1, \ldots, n$. Then the moment generating function for $S_R = \mathcal{R}_1 + \cdots + \mathcal{R}_n$ is

$$\Gamma(s, \Pi, R) = \prod_{i=1}^{n} \gamma_i(s, \pi_i, r_i), \tag{4.59}$$

and the Chernoff bound says that

$$\Pr\{S_n \leq \zeta\} \leq \min_{s \geq 0} \{\Gamma(-s, \Pi, R) e^{s\zeta}\}. \tag{4.60}$$

Thus if we define

$$G^{X}(\Pi, \zeta) = \min_{\|R\|^2=1} \min_{s \geq 0} \left\{ \Gamma(-s, \Pi, R) e^{s\zeta} \right\} \quad \text{and} \tag{4.61}$$

$$R^{X}(\Pi, \zeta) = \arg_{R} \min_{\|R\|^2=1} \min_{s \geq 0} \left\{ \Gamma(-s, \Pi, R) e^{s\zeta} \right\}, \tag{4.62}$$

we have the following theorem.

Theorem 4.9. $P(\Pi, \infty) = P^{*}(\Pi, \infty) = G^{*}(\Pi, \sqrt{v}) \leq G^{X}(\Pi, \sqrt{v}).$

The function $G^{X}(\Pi, \sqrt{v}) = G^{X}(\Pi, \sqrt{k-1})$ is our main tool, since it is (a) relatively easy to calculate, and (b) a tight upper bound on $P(\Pi, \infty)$, at least when $P(\Pi, \infty)$ is small. Furthermore, the matrix $R^{X}(\Pi, \sqrt{k-1})$, when appropriately scaled and quantized, represents a near-optimal choice for the multiplicity matrix for large values of the cost. In the next section, we derive key equations which form the heart of the algorithm used to find the near-optimum multiplicity matrices.

4.7.3 The Lagrangian

In this section, we will focus on finding the optimum matrix $\mathbb{X} = (X_i(\beta))$ with a finite cost γ and with L^2 and D' defined as in (4.51). As seen in the previous section, the case of an optimum infinite-cost multiplicity matrix is the special case with $L^2 = 1$ and $D' = \sqrt{v}$. The problem of finding the optimum matrix, $\mathbb{X}^{*}$, in (4.53) could be reformulated as the constrained optimization problem,

$$\min \left(sD' + \sum_{i=1}^{n} \ln \phi_i(-s, \pi_i, X_i) \right) \tag{4.63}$$

subject to

$$s \geq 0$$

$$\|\mathbb{X}\|^2 = L^2 = 2\gamma + \frac{1}{4}nq.$$

Define the Lagrangian,

$$\mathcal{L}(s, \mathbb{X}, \lambda) = sD' + \sum_{i=1}^{n} \ln \phi_i(-s, \pi_i, X_i) + \frac{\lambda}{2} \left(\|\mathbb{X}\|^2 - L^2 \right).$$

It is required to solve for $s^*, \mathbb{X}^*$, and λ^* that satisfy

$$\left. \frac{\partial \mathcal{L}}{\partial \lambda} \right|_{\lambda = \lambda^*} = 0, \quad \left. \frac{\partial \mathcal{L}}{\partial s} \right|_{s = s^*} = 0 \quad \text{and} \quad \left. \frac{\partial \mathcal{L}}{\partial X_i(\beta)} \right|_{\mathbb{X} = \mathbb{X}^*} = 0.$$

If the optimization for s results in a negative value for s^*, then this value is discarded and s^* is taken to be at the boundary, i.e., $s^* = 0$. (This may be the case at low signal-to-noise ratios when the matrix Π has a random-like structure.) The corresponding optimized multiplicity matrix X^* is calculated by optimizing for X.

Since $D' = D_v(\gamma) + n/2$ and $\gamma = (\|\mathbb{X}\|^2 - \frac{nq}{4})/2$, then D' is a function of $\mathbb{X}$. Since $D_v(\gamma)$ is actually a discrete function, then it could not be differentiated, however it is well approximated by the continuous upper bound in (4.8),

$$\frac{\partial D'}{\partial X_i(\beta)} \approx \left(\frac{\sqrt{v}}{\sqrt{\|\mathbb{X}\|^2 - \frac{nq}{4}}} - \frac{v^{3/2}}{8 \left(\|\mathbb{X}\|^2 - \frac{nq}{4} \right)^{3/2}} \right) X_i(\beta) = \psi(\|\mathbb{X}\|^2) X_i(\beta).$$

In fact the term $\psi(\|\mathbb{X}\|^2)$ will cancel while solving for $\mathbb{X}^*$ below. Solving for $\mathbb{X}^*$ and s^*;

$$\left. \frac{\partial \mathcal{L}}{\partial \lambda} \right|_{\lambda = \lambda^*} = 0 \implies \|\mathbb{X}\|^2 = L^2, \tag{4.64}$$

$$\left. \frac{\partial \mathcal{L}}{\partial s} \right|_{s = s^*} = D' - \sum_{i=1}^{n} \left(\frac{\sum_{\beta \in \mathbb{F}_q} X_i(\beta) \pi_i(\beta) e^{-sX_i(\beta)}}{\phi_i(-s, \pi_i, X_i)} \right) \Bigg|_{s = s^*} = 0, \tag{4.65}$$

$$\left. \frac{\partial \mathcal{L}}{\partial X_i(\beta)} \right|_{\mathbb{X} = \mathbb{X}^*} = s\psi(\|\mathbb{X}\|^2) X_i(\beta) - s \frac{\pi_i(\beta) e^{-sX_i(\beta)}}{\phi_i(-s, \pi_i, X_i)} + \lambda X_i(\beta) \Bigg|_{\mathbb{X} = \mathbb{X}^*} = 0. \tag{4.66}$$

Multiplying (4.66) by $X_i(\beta)$, summing over $\beta \in \mathbb{F}_q$ and then summing over i, we get

$$s\psi(\|\mathbb{X}\|^2)\|\mathbb{X}\|^2 - s \sum_{i=1}^{n} \left(\frac{\sum_{\beta \in \mathbb{F}_q} X_i(\beta) \pi_i(\beta) e^{-sX_i(\beta)}}{\phi_i(-s, \pi_i, X_i)} \right) + \lambda \|\mathbb{X}\|^2 \Bigg|_{\mathbb{X} = \mathbb{X}^*} = 0. \tag{4.67}$$

Substituting (4.64) and rearranging;

$$\lambda = s \left(\frac{1}{L^2} \sum_{i=1}^{n} \left(\frac{\sum_{\beta \in \mathbb{F}_q} X_i(\beta) \pi_i(\beta) e^{-s X_i(\beta)}}{\phi_i(-s, \pi_i, X_i)} \right) - \psi(L^2) \right). \tag{4.68}$$

Substituting back in (4.66) we reach the following equation,

$$\frac{X_i(\beta)}{L^2} \sum_{i=1}^{n} \left(\frac{\sum_{\beta \in \mathbb{F}_q} X_i(\beta) \pi_i(\beta) e^{-s X_i(\beta)}}{\phi_i(-s, \pi_i, X_i)} \right) - \frac{\pi_i(\beta) e^{-s X_i(\beta)}}{\phi_i(-s, \pi_i, X_i)} \Bigg|_{\mathbf{X} = \mathbf{X}^*} = 0. \tag{4.69}$$

If $s = s^*$, then this equation reduces to

$$\frac{D'}{L^2} X_i(\beta) - \frac{\pi_i(\beta) e^{-s^* X_i(\beta)}}{\sum_{\beta \in \mathbb{F}_q} \pi_i(\beta) e^{-s^* X_i(\beta)}} \Bigg|_{\mathbf{X} = \mathbf{X}^*} = 0. \tag{4.70}$$

In summary, the optimization problem is reduced to finding s^* and $\mathbf{X}^*$ which are the solutions for equations (4.65) and (4.69) (or (4.70)), respectively.

4.7.4   Convexity

In this section, we show that the optimized Lagrangian, $\mathcal{L}^*(s, \mathbf{X}) = \mathcal{L}(s, \mathbf{X}, \lambda^*)$, is convex in both s and $\mathbf{X}$. Thus an iterative algorithm that will minimize $\mathcal{L}^*(s, \mathbf{X})$ could be developed. Specifically we show that for a given multiplicity matrix X', the optimized Lagrangian is convex in the parameter s, and for a given s (at $s = s*$), the optimized Lagrangian is convex in the nq variables which are the components of the multiplicity matrix $\mathbf{X}$. Let

$$\mathcal{L}_s(s) \triangleq \mathcal{L}^*(s, \mathbf{X})|_{\mathbf{X} = X'} \tag{4.71}$$

$$\mathcal{L}_{\mathbf{X}}(\mathbf{X}) \triangleq \mathcal{L}^*(s, \mathbf{X})|_{s=s^*}. \tag{4.72}$$

$\mathcal{L}_s(s)$ is Convex in s

The gradient of $\mathcal{L}_s(s)$ is defined to be $G_s(s) = \frac{\partial \mathcal{L}_s(s)}{\partial s}$ and is given by (4.65).

The second derivative of $\mathcal{L}_s(s)$ with respect to s is

$$\frac{\partial^2 \mathcal{L}_s(s)}{\partial s^2} = \sum_{i=1}^{n} \left(\frac{\sum_{\beta \in \mathbb{F}_q} X_i^2(\beta) \pi_i(\beta) e^{-sX_i(\beta)}}{\sum_{\beta \in \mathbb{F}_q} \pi_i(\beta) e^{-sX_i(\beta)}} - \left(\frac{\sum_{\beta \in \mathbb{F}_q} X_i(\beta) \pi_i(\beta) e^{-sX_i(\beta)}}{\sum_{\beta \in \mathbb{F}_q} \pi_i(\beta) e^{-sX_i(\beta)}} \right)^2 \right).$$

Define the $q \times 1$ -dimensional vectors Λ_i and Θ_i such that

$$\Lambda_i = \left\{ X_i(\beta) \sqrt{\pi_i(\beta)} e^{-sX_i(\beta)/2} \right\} \text{ and } \Theta_i = \left\{ \sqrt{\pi_i(\beta)} e^{-sX_i(\beta)/2} \right\} \text{ for } \beta \in \mathbb{F}_q,$$

then the second derivative of $\mathcal{L}_s(s)$ with respect to s is reformulated as

$$H_s = \frac{\partial^2 \mathcal{L}_s(s)}{\partial s^2} = \sum_{i=1}^{n} \left(\frac{\|\Lambda_i\|^2 \|\Theta_i\|^2 - (\Lambda_i^T \Theta_i)^2}{\|\Theta_i\|^4} \right),$$

where for any vector $\boldsymbol{x}$, $\|\boldsymbol{x}\| = \left(\boldsymbol{x}^T \boldsymbol{x}\right)^{1/2}$ is the Euclidean norm of $\boldsymbol{x}$. By the Cauchy Schwartz inequality

$$\|\Lambda_i\| \|\Theta_i\| \geq (\|\Lambda_i^T \Theta_i\|)\|_1,$$

where $\|.\|_1$ is absolute value and $(.)^T$ is the vector transposed, with equality iff there exists an $\alpha \geq 0$ such that $\Lambda_i = \alpha \Theta_i$. Thus $H_s \geq 0$, which implies that $\mathcal{L}_s(s)$ is *convex*. In fact, $H_s = 0$ iff for each $i = 1, \ldots, n$, $X_i(\beta) = \alpha_i$ where $\alpha_i \geq 0$ for all $\beta \in \mathbb{F}_q$. Since $X_i(\beta)$ is a function of $\pi_i(\beta)$, then this implies that for each i, $\pi_i(\beta) = 1/q$. This would imply that all symbols $\beta \in \mathbb{F}_q$ are equally likely given the received symbol. At reasonable operating conditions, such a condition does not occur for all $i = 1, \ldots, n$, as it is equivalent to receiving all n symbols of the codeword in error. So in general, $H_s > 0$ and $\mathcal{L}_s(s)$ is *strongly convex* in s.

$\mathcal{L}_{\mathbb{X}}(\mathbb{X})$ is Convex in $\mathbb{X}$

Define the qn-dimensional vector

$$\bar{X} = \{X_1(\beta_1), \ldots, X_1(\beta_q), \ldots, X_n(\beta_1), \ldots, X_n(\beta_q)\}.$$

So the gradient of $\mathcal{L}_{\mathbb{X}}(\mathbb{X})$ is defined by the $(qn \times 1)$-dimensional vector,

$$G_X = \left\{ G_{X_1(\beta_1)}, \ldots, G_{X_1(\beta_q)}, \ldots, G_{X_n(\beta_1)}, \ldots, G_{X_n(\beta_q)} \right\},$$

where

$$G_{X_i(\beta)} = \frac{\partial \mathcal{L}_{\mathbb{X}}(\mathbb{X})}{\partial X_i(\beta)} = s^* \left(\frac{D'}{L^2} X_i(\beta) - \frac{\pi_i(\beta) e^{-s^* X_i(\beta)}}{\sum_{\beta \in \mathbb{F}_q} \pi_i(\beta) e^{-s^* X_i(\beta)}} \right). \tag{4.73}$$

The second derivatives are given by

$$\frac{1}{s^*} \frac{\partial^2 \mathcal{L}_{\mathbb{X}}(\mathbb{X})}{\partial X_i^2(\beta)} = \frac{D'}{L^2} + s^* \pi_i(\beta) e^{-s^* X_i(\beta)} \frac{\left(\sum_{\beta \in \mathbb{F}_q} \pi_i(\beta) e^{-s^* X_i(\beta)} \right) - \pi_i(\beta) e^{-s^* X_i(\beta)}}{\left(\sum_{\beta \in \mathbb{F}_q} \pi_i(\beta) e^{-s^* X_i(\beta)} \right)^2},$$

$$\frac{1}{s^*} \frac{\partial^2 \mathcal{L}_{\mathbb{X}}(\mathbb{X})}{\partial X_i(\beta_1) \partial X_i(\beta_2)} \bigg|_{\beta_1 \neq \beta_2} = -s^* \frac{\pi_i(\beta_1) \pi_i(\beta_2) e^{-s^* (X_i(\beta_1) + X_i(\beta_2))}}{\left(\sum_{\beta \in \mathbb{F}_q} \pi_i(\beta) e^{-s^* X_i(\beta)} \right)^2}, \quad \text{and}$$

$$\frac{1}{s^*} \frac{\partial^2 \mathcal{L}_{\mathbb{X}}(\mathbb{X})}{\partial X_i(\beta_1) \partial X_j(\beta_2)} \bigg|_{\beta_1 \neq \beta_2, i \neq j} = 0.$$

Define the $q \times q$ matrix, H_{X_i}, such that for $a, b = 1, 2, \ldots, q$,

$$[H_{X_i}]_{a,b} = \frac{\partial^2 \mathcal{L}_{\mathbb{X}}(\mathbb{X})}{\partial X_i(\beta_a) \partial X_i(\beta_b)},$$

then using the above second-order derivatives

$$H_{X_i} = s^* \left(\frac{D'}{L^2} I_q + \frac{s^*}{(J^T z_i)^2} \left((J^T z_i) \operatorname{Diag}(z_i) - z_i z_i^T \right) \right),$$

where $z_i = \left\{ \pi_i(\beta_a) e^{-s^* X_i(\beta_a)}, \ a = 1, \ldots, q \right\}$ is a $(q \times 1)$ vector, J is the all-ones q vector and $\operatorname{Diag}(z)$ is the diagonal matrix with the elements of z on the diagonal. The Hessian of $\mathcal{L}_{\mathbb{X}}(\mathbb{X})$ defined by

$$[H_X]_{a,b} = \frac{\partial^2 \mathcal{L}_{\mathbb{X}}(\mathbb{X})}{\partial \bar{X}(a) \partial \bar{X}(b)}$$

is thus given by the block diagonal matrix

$$H_X = \operatorname{Diag}(H_{X_1}, H_{X_2}, \ldots, H_{X_n}). \tag{4.74}$$

Let v_i be any q vector,

$$\Psi_i = \left\{ \sqrt{z_i(1)}, \ldots, \sqrt{z_i(q)} \right\}^T \quad \text{and} \quad \Phi_i = \left\{ v_i(1)\sqrt{z_i(1)}, \ldots, v_i(q)\sqrt{z_i(q)} \right\}^T,$$

then

$$v_i^T H_{X_i} v_i = s^* \left(\frac{D'}{L^2} v_i^T v_i + \frac{s^*}{(J^T z_i)^2} \left((\Psi_i^T \Psi_i)(\Phi_i^T \Phi_i) - (\Psi_i^T \Phi_i)^2 \right) \right), \tag{4.75}$$

By the Cauchy-Schwartz inequality,

$$(\Psi_i^T \Psi_i)(\Phi_i^T \Phi_i) - (\Psi_i^T \Phi_i)^2 \geq 0,$$

and by substituting in (4.75) it follows that

$$v_i^T H_{X_i} v_i \geq \frac{s^* D'}{L^2} v_i^T v_i \geq 0, \tag{4.76}$$

where the last inequality is due to the fact that $s^* \geq 0$ and $v_i^T v_i = \|v_i\|^2 \geq 0$ for any vector v_i. If $s^* > 0$, then $v_i^T H_{X_i} v_i > 0$ for any nonzero vector v_i which implies that H_{X_i} is *positive definite*. Let $v = \{v_1^T, v_2^T, \ldots, v_n^T\}^T$ be an arbitrary qn vector, then from (4.76) and (4.74), it follows that

$$v^T H_X v = \sum_{i=1}^{n} v_i^T H_{X_i} v_i \geq 0,$$

which proves that $\mathcal{L}_X(\mathbf{X})$ is convex. Generally, $s^* > 0$ which would imply that H_X is *positive definite* and thus $\mathcal{L}_X(\mathbf{X})$ is *strongly convex*. In this analysis, we assumed that $s = s^*$ since we will optimize for s and then for $\mathbf{X}$. However, for another $s \geq 0$, the term D' in (4.73) could be treated as another positive quantity and the analysis holds.

4.7.5 Iterative Algorithm

The proposed iterative algorithm for finding $\mathbf{X}^* = (X_i(\beta))$, and thus the optimum multiplicity matrix, could be summarized as follows:

Algorithm 4.1. *Let s^j and $\mathbb{X}^j = (X_i^j(\beta))$ be the values of s and $\mathbb{X}$ at the jth iteration respectively. $\epsilon \approx 10^{-5}$ is a small number greater than zero.*

Initialize $\mathbb{X}^o = \frac{L^2}{D'}\Pi,\ \ s^o = 0.1 * \frac{D'}{L^2}\ \ and\ \ j = 0.$

Do

    $j := j + 1$

    I. **Solve** *for s^j, (4.65),*

$$\nabla_s \left(\mathcal{L}^*(s, \mathbb{X}^{j-1}) \right) = \left. \frac{\partial \mathcal{L}^*(s, \mathbb{X}^{j-1})}{\partial s} \right|_{s=s^j} = 0$$

If s^j is negative then set s^j to be zero.

    II. **Solve** *for $\mathbb{X}^j$, (4.69),*

$$\nabla_{\mathbb{X}} \left(\mathcal{L}^*(s^j, \mathbb{X}) \right) = \left. \left\{ \frac{\partial \mathcal{L}^*(s^j, \mathbb{X})}{\partial X_i^j(\beta)}, i = 1, \ldots, n,\ \beta \in \mathbb{F}_q \right\} \right|_{\mathbb{X}=\mathbb{X}^j} = 0$$

While

$$\left\| \frac{s^j - s^{j-1}}{s^{j-1}} \right\|_1 \leq \epsilon.$$

For the case of finite costs, the optimized integer multiplicity matrix, $M = (m_i(\beta))$ is found from the optimized matrix $\mathbb{X}^* = (X_i^*(\beta))$ by the inverse transformation,

$$m_i(\beta) = \text{Round} \left\{ \max \left\{ 0, X_i^*(\beta) - 0.5 \right\} \right\}, \tag{4.77}$$

where $\text{Round}\,\{\}$ *is the rounding to the nearest integer.*

4.7.6   Implementation Issues

In our implementation and for the simulation results in this chapter, we replace the command **Solve** by a Newton-type algorithm. Other algorithms such as the gradient descent algorithm, which is less computationally complex, were also tested. However, the Newton algorithm described in Appendix A achieved the best results. Given that the complexity of Newton's algorithm can be cubic in the number of optimized variables and we the Chernoff algorithm is an optimization in qn variables, the entries of $\mathbb{X}$, it is computationally expensive. However, the computational complexity could be reduced dramatically by observing that the entries of each column in Π, π_i, sum to one, and that for reasonable operating signal-to-noise ratios (SNRs) only a small fraction of them have a relevant value while the rest tend to be negligible or zero. Thus, in optimizing for $\mathbb{X}$ only the elements $X_i(\beta)$ corresponding to elements $\pi_i(\beta)$ above a certain threshold are considered for optimization while the others are set to zero. Practically, this threshold could be set to 10^{-6} or 10^{-7}. This implies that the complexity of our algorithm decreases with an increase in the operating SNR, which is usually the case for operating conditions. Another approach, which might be less reliable but whose complexity is independent of the SNR, is to optimize only for the largest ϵ entries in each column rendering the number of optimized variables to be $n\epsilon$.

4.8 Numerical Results

In this section we will refer to our method as the Chernoff method. The Gaussian approximation of [83] is referred to as the Gauss method and the Koetter-Vardy algorithm, (4.21), as KV. A hard decision bounded minimum distance decoder, as the Berlekamp-Massey algorithm, is referred to as BM. It is to be noted that we used the condition of (4.4) to test if the transmitted codeword is on the GS generated list for all ASD algorithms compared. If the sufficient condition is satisfied then a decoding success is signaled. This is somehow justified by the fact that, on average, the list size is one [77]. If the GS generated list is empty, then it is most likely that the sufficient condition will not be satisfied, and a decoder error is signaled. In a real time implementation, if more than one codeword is on the generated list, then the most reliable codeword (with respect to the soft output from the channel) is chosen as the decoder output.

To test our theories, we simulated the performance of the $(15, 11)$ RS code over the finite field $\mathbb{F}_q$ of 16 elements, $\mathbb{F}_{16}$, on an additive white Gaussian noise (AWGN) channel. These results are shown in Figure 4.2 and Figure 4.3 for the cases of binary phase shift keying (BPSK) and 16-ary phase shift keying (PSK) modulation schemes respectively.

We see that the Chernoff technique shows a marked superiority when compared to

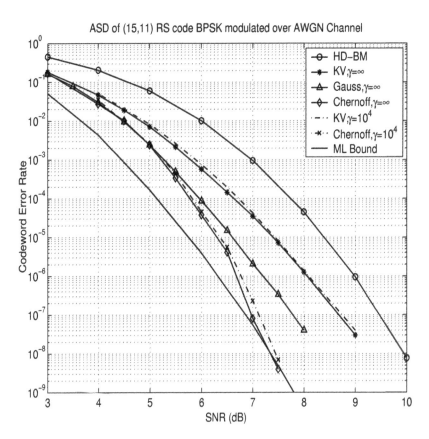

Figure 4.2: Performance of ASD algorithms when decoding an (15, 11) RS code BPSK modulated over an AWGN channel, for both finite and infinite interpolation costs. Their performance is also compared to an averaged upper bound on the performance of the ML decoder.

the KV technique, for both finite and infinite cost matrices. For BPSK modulation, infinite cost γ, and an error rate of 4×10^{-8}, our algorithm has about 0.9 dB, 1.8 dB and 2.5 dB coding gains over the Gauss, KV and BM algorithms respectively. Simulation results for a finite cost of 10^4 also show the potential of our algorithm over previously proposed ones. A tight averaged upper bound on the maximum-likelihood error probability (Section 2.4) is also plotted. Since it is the binary image of the RS code which is modulated and transmitted over the channel, and the binary image is not unique but depends on the basis used to represent the symbols in $\mathbb{F}_{16}$ as bits, this bound was derived by averaging over all possible binary images over an RS code. By comparing with actual simulations for maximum-likelihood decoding of the $(15, 11)$ RS code over an AWGN channel this bound was shown to be tight (Chapter 2). Our algorithm has a near-ML performance at high signal-to-noise ratios.

Similarly, for the case of 16-ary PSK, the Chernoff algorithm has about 2.6 dB gain over the BM algorithm at a codeword error rate of 10^{-7}. The performance gain over KV is about 1.7 dB at an error rate of 10^{-6}.

Numerical results for ASD decoding of the $(31, 25)$ RS code over $\mathbb{F}_{32}$ BPSK modulated over AWGN channel are shown in Figure 4.4. As seen the Chernoff algorithm has up to 2 dB gain over the hard-decision BM algorithm. The coding gain over the KV algorithm and the Gaussian approximation increases at the tail of error probability. The averaged bound on the ML error probability is also plotted. It is observed that that at high SNRs, our algorithm is near optimal.

To demonstrate the convergence of our proposed algorithm, we plot the value of s^j, (see Algorithm 4.1), versus the iteration number j for a fixed value of SNR. This is shown in Figure 4.5 for a randomly transmitted $(15, 11)$ RS codeword and BPSK modulation with an SNR of 6 dB. The average codeword error rate is plotted in Figure 4.6 versus the number of iterations at a SNR of 5.5 dB. These figures demonstrate the fast convergence of the algorithm in terms of the number of (global) iterations.

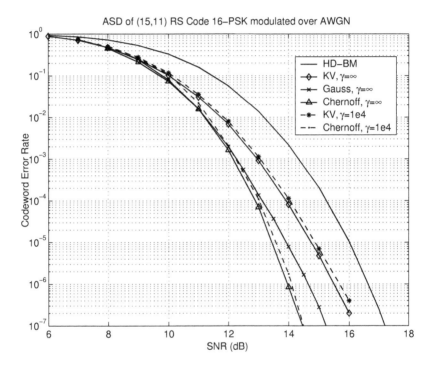

Figure 4.3: Performance curves for decoding an $(15, 11)$ RS code, 16-PSK modulated over an AWGN channel, using different ASD algorithms.

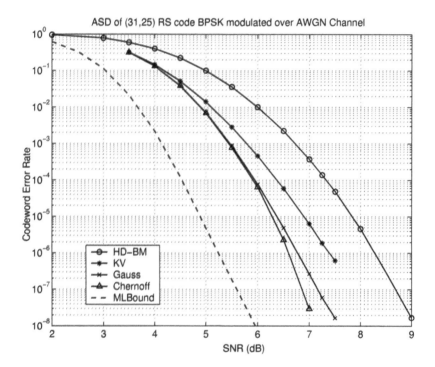

Figure 4.4: An $(31, 25)$ RS code is BPSK modulated over an AWGN channel. ASD algorithms are compared at infinite interpolation costs. The Chernoff algorithm has a better performance than the Gauss and KV algorithms. The performance curve of a bounded minimum distance decoder and an averaged upper bound on the performance of the ML decoder are also plotted.

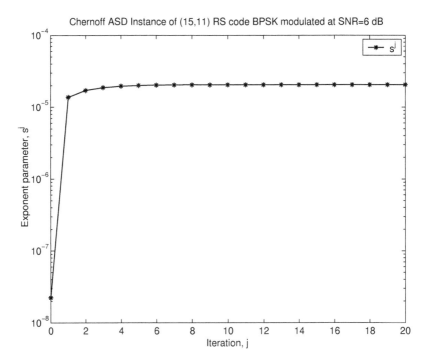

Figure 4.5: Convergence of the Chernoff bound algorithm at an SNR of 6 dB.
A decoding instance of the $(15, 11)$ RS code, BPSK modulated over an AWGN channel
at a fixed SNR of 6 dB, using Chernoff ASD. The convergence of the algorithm is
conveyed by the fast adaptation of the exponential parameter s^j.

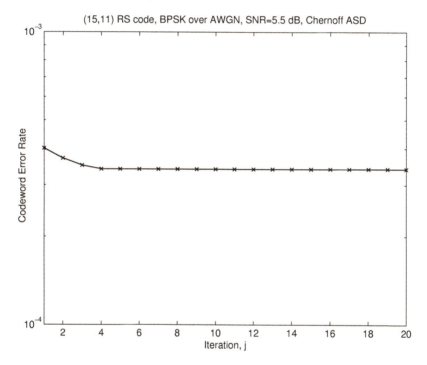

Figure 4.6: Convergence of the error probability of the Chernoff bound algorithm. The convergence is demonstrated by plotting the average codeword error probability versus the number of iterations at a fixed SNR of 5.5 dB.

4.9   Discussion

The performance gains of our algorithm over that of the Gaussian approximation, by
Parvaresh and Vardy [83], could be reasoned by observing that the Gaussian approximation finds the multiplicity matrix of infinite cost that minimizes the error probability
assuming that the score has a Gaussian distribution. It could be shown that this is
equivalent to minimizing the Chebychev bound [89, 118] on the error probability assuming that the score is symmetrically distributed around its mean; By the Chebyshev
bound (c.f (4.49)),

$$\Pr\{S_M \leq \delta\} \leq \frac{\sigma_S^2}{2(\delta - \mu_S)^2}, \tag{4.78}$$

where μ_S and σ_S^2 are the mean and variance of the score and it is assumed that $\mu_S - \delta \geq$
0. The expectation is given by

$$\mu_S = E\{S_M\} = \sum_{i=1}^{n} E\{S_i\} = \sum_{i=1}^{n} \sum_{\beta \in \mathbb{F}_q} \pi_i(\beta)\, m_i(\beta), \tag{4.79}$$

where $\Pi = (\pi_i(\beta))$ is the reliability matrix. With the assumption that all the random
variables S_i are independent,

$$\sigma_S = \sum_{i=1}^{n} \sum_{\beta \in \mathbb{F}_q} \pi_i(\beta)\, m_i^2(\beta) - \sum_{i=1}^{n} \left(\sum_{\beta \in \mathbb{F}_q} \pi_i(\beta)\, m_i(\beta) \right)^2. \tag{4.80}$$

The minimizing multiplicities are found by differentiating the bound (4.78) with respect
to $m_i(\beta)$ and equating to zero.

It is well known that the Chernoff bound is a tighter upper bound than the Chebychev bound (c.f., [89, 118]). Further more, no assumptions about the distribution of
the score is made in deriving our algorithm.

It is observed that the coding gains of the Chernoff algorithm, developed in this
chapter, over other ASD algorithms increases as the SNR increases and approaches
that of the ML bound. This somehow proves the conjecture that our algorithm is
optimal at the tail of error probability. The reasoning behind that is the fact that the

Chernoff bound, in general, is an exponentially tight upper bound at the tail of error probability and closely approximates the true error probability. In another way, this shows the potential of using the Chernoff algorithm in favorable operating conditions.

4.10 Conclusion

The goal of this chapter was to find the ultimate capabilities of algebraic soft decoding of Reed-Solomon codes. Since the performance of ASD depends mainly on the interpolation multiplicities assigned, we explored a novel multiplicity assignment algorithm that results in an improved performance. The multiplicity assignment algorithm proposed aims at directly minimizing the decoding error probability. Reasonable approximations and relaxations were made to simplify the problem. However, since the actual error probability is relatively hard to compute, we aimed at finding the multiplicity matrix that will minimize an upper bound (the Chernoff bound) on the error probability. We explore the cases of both finite and infinite cost multiplicity matrices. The problem is formulated as a constrained optimization problem and an iterative algorithm is developed that will find the optimum multiplicity matrix. Numerical results show that our algorithm is superior to other multiplicity assignment algorithms found in the literature.

Chapter 5

Iterative Algebraic Soft-Decision Decoding of Reed-Solomon Codes

When I have fully decided that a result is worth getting I go ahead of it
and make trial after trial until it comes.

—Thomas A. Edison

As we mentioned in Chapter 4, the performance of algebraic soft-decision decoding of Reed-Solomon codes depends on the scheme used to assign multiplicities for the Guruswami-Sudan algorithm. While searching for the optimum multiplicity matrix, we have proposed a multiplicity assignment algorithm, based on the Chernoff bound, that has best performance when compared to other previously proposed multiplicity assignment algorithms. The gap to the maximum-likelihood performance hinted at the possible existence of even better algebraic soft-decision decoding algorithms. In this chapter, we develop an algebraic soft-decision list-decoding algorithm based on the idea that belief propagation-based algorithms could be deployed to improve the reliability of the symbols that is then utilized by an interpolation multiplicity assignment algorithm.

Conventional message passing algorithms, when applied on RS codes, may not result in a good performance due to the dense nature of the associated parity check matrices. Jiang and Narayanan (JN) developed an iterative algorithm based on belief propagation for soft decoding of RS codes [64, 63]. This algorithm compares favorably with other soft-decision decoding algorithms for RS codes (c.f., [88]) and is a major step towards message passing decoding algorithms for RS codes. In the JN algorithm, belief propagation is run on an *adapted* parity check matrix where the columns in the parity-check matrix corresponding to the least reliable independent bits are reduced to an identity submatrix [64, 63]. The order statistics decoding algorithm by Fossorier and Lin [42] also sorts the received bits with respect to their reliability information and reduces the columns in the generator matrix corresponding to the most reliable bits to an identity submatrix. This matrix is then used to generate (permuted) codewords using the most reliable bits. Other soft-decoding algorithms for RS codes include the generalized minimum distance (GMD) decoding algorithm introduced by Forney [41], the Chase II algorithm [18], the combined Chase II-GMD algorithm [103], successive erasure-error decoding [60] as well as code decomposition [53].

For a brief review of the GS algorithm and algebraic soft-decision decoding, in particular the Koetter-Vardy algorithm, we refer the reader to Section 4.2 and Section 4.5 in the previous chapter. An outline of this chapter is as follows. In Section 5.1, we introduce some notation and describe the technique we used to derive the binary images of Reed-Solomon codes. The JN algorithm is explained in the context of this chapter in Section 5.2. Some modifications to the JN algorithm are introduced in Section 5.3. One of the main contributions in this chapter, the iterative algebraic soft-decision list-decoding algorithm, is presented in Section 5.4. Another main contribution, a low complexity algorithm based on the JN algorithm, is presented in Section 5.5. Some discussions as well as some numerical results are presented in section 5.6. Finally, we conclude the chapter in Section 5.7.

5.1   Preliminaries

As in Chapter 4, $\boldsymbol{d} = [d_0, d_1, \ldots, d_{k-1}]$ will denote a k-dimensional vector over $\mathbb{F}_q$ where $\mathbb{F}_q$ is the finite field of q elements. $\mathcal{C}$ will denote an (n, k, d) RS code. An (n, k, d) RS codeword $\boldsymbol{u} = [u_0, u_1, \ldots, u_{n-1}]$ could be generated by evaluating a data polynomial $\mathbb{D}(X) = \sum_{i=0}^{k-1} d_i X^i$, of degree $k - 1$, at n elements of the field composing a set, called the support set of the code;

$$\boldsymbol{u} = [\mathbb{D}(1), \mathbb{D}(\alpha^1), \ldots, \mathbb{D}(\alpha^{n-1})], \tag{5.1}$$

where α is the primitive element of the field and $n = q - 1$. The set $S = \{1, \alpha, \alpha^2, \ldots, \alpha^{n-1}\}$ is called the support set of the code and is vital for the operation of the Guruswami-Sudan algorithm.

Lemma 5.1. *The polynomial* $\mathbb{U}(X) = \sum_{i=0}^{n-1} u_i X^i$ *associated with a codeword* $\boldsymbol{u}$ *generated as in (5.1) has* $\alpha, \alpha^2, \ldots, \alpha^{n-k}$ *as zeros.*

Proof. Let $\boldsymbol{d}'$ be the vector $\boldsymbol{d}$ padded with $(n-k)$ zeros such that

$$d_i' = \begin{cases} d_i \text{ for } i \le k-1 \\ 0 \text{ for } k \le i \le n-1. \end{cases} \tag{5.2}$$

It follows from (5.1) that $u_j = \sum_{i=0}^{n-1} d_i' \alpha^{ij}$. Thus $\boldsymbol{u}$ is the discrete Fourier transform (DFT) of $\boldsymbol{d}'$ [74]. The inverse DFT of $\boldsymbol{u}$ is given by

$$d_j' = \sum_{i=0}^{n-1} u_i \alpha^{-ij} = \mathbb{U}(\alpha^{-j}) = \mathbb{U}(\alpha^{n-j}). \tag{5.3}$$

Substituting (5.2) in (5.3), we conclude that $\mathbb{U}(\alpha^i) = 0$ for $i = 1, 2, \ldots, n-k$. □

It follows that

$$\sum_{i=0}^{n-1} u_i \alpha^{ij} = 0 \text{ for } j = 1, 2, \ldots, n-k, \tag{5.4}$$

and a valid parity check matrix $\mathcal{H}$, such that $\mathcal{H}\boldsymbol{u}^T = 0$, is [75]

$$\mathcal{H} = \begin{bmatrix} 1 & \alpha & \cdots & \alpha^{n-1} \\ 1 & \alpha^2 & \cdots & \alpha^{2(n-1)} \\ \vdots & \vdots & \cdots & \vdots \\ 1 & \alpha^{n-k} & \cdots & \alpha^{(n-k)(n-1)} \end{bmatrix}. \tag{5.5}$$

5.1.1   A Binary Image of the Reed-Solomon Code

In many cases, it is the *binary image* of RS codes which is modulated and transmitted over the channel. We show here a valid binary representation of a RS code and it corresponding parity check matrix. Let $\mathbb{P}(X) = a_o + a_1 X + a_{m-1} X^{m-1} + X^m$ be a primitive polynomial in $\mathbb{F}_2[X]$. Let α be a root of $\mathbb{P}(X)$, then α is a primitive element in $\mathbb{F}_{2^m}$. The companion matrix of $\mathbb{P}(X)$ is given by the $m \times m$ matrix

$$C = \left[\begin{array}{ccc|c} 0 \dots & & 0 & a_o \\ & & & a_1 \\ & I_{m-1} & & \vdots \\ & & & a_{m-1} \end{array} \right], \tag{5.6}$$

where I_m is the $m \times m$ identity matrix [59]. The characteristic polynomial of this matrix satisfies

$$\det(C - I_m X) = \mathbb{P}(X).$$

Representing the primitive element, α, by its binary companion matrix C, the mapping $\alpha^i \leftrightarrow C^i$, $\{i = 0, 1, 2, \dots\}$ induces a field isomorphism. So every element in the parity check matrix of (5.5) can be replaced with an $m \times m$ matrix resulting in a binary parity check matrix H of size $(n-k)m \times nm$. Also, any element, $\beta \in \mathbb{F}_{2^m}$, has an m-tuple representation $\{\beta_0, \beta_1, \dots, \beta_{m-1}\}$ where

$$\beta = \beta_0 + \beta_1 \alpha + \dots + \beta_{m-1} \alpha^{m-1}, \quad \beta_i \in \mathbb{F}_2 \tag{5.7}$$

and α is the primitive element of the field. Let $\beta = \alpha^i$ then it follows that

$$C^j \{\beta_0, \beta_1, \dots, \beta_{m-1}\}^T \leftrightarrow \alpha^{ij}$$

where the matrix multiplication is done in $\mathbb{F}_2$. The binary image of a codeword $\boldsymbol{u}$ is given by the nm tuple $\boldsymbol{u_b}$ where

$$\boldsymbol{u_b} = [u_{0,0}, u_{0,1}, \dots, u_{0,m-1}, \dots, u_{n-1,0}, u_{n-1,1}, \dots, u_{n-1,m-1}].$$

Such a mapping results in $H\boldsymbol{u_b}^T = 0$. The redundancy of the code's binary image is $\tilde{r}$ where $\tilde{r} = \tilde{n} - \tilde{k}$, $\tilde{n} = mn$ and $\tilde{k} = mk$.

Throughout this chapter, the received vector will be denoted by $\boldsymbol{y} = \boldsymbol{x} + \boldsymbol{\eta}$, where $\boldsymbol{x} = 1 - 2\boldsymbol{u_b}$ is the BPSK modulation of a codeword $\boldsymbol{u}$ and $\boldsymbol{\eta}$ is the AWGN vector with variance σ^2. The channel log likelihood ratios (LLRs) are given by $\boldsymbol{\Lambda}^{ch} = 2\boldsymbol{y}/\sigma^2$. In concatenated coding systems, where the RS code is implemented as an outer code, the "channel" LLRs will be the soft output of an inner decoder such as the BCJR algorithm [7], the soft output Viterbi algorithm (SOVA) [112] or another BP decoder.

5.2   Adaptive Belief Propagation

Gallager devised an iterative algorithm for decoding his low-density parity check (LDPC) codes [45]. This algorithm was the first appearance in the literature of what we now call belief propagation (BP). Recall that H is the parity check matrix associated with the binary image of the RS code. It has $\tilde{r}$ rows corresponding to the check nodes and $\tilde{n}$ columns corresponding to the variable nodes (transmitted bits). $H_{i,j}$ will denote the element in the ith row and jth column of H. Define the sets, $J(i) \triangleq \{j \mid H_{i,j} = 1\}$ and $I(j) \triangleq \{i \mid H_{i,j} = 1\}$. Define $Q_{i,j}$ to be the log-likelihood ratio (LLR) of the jth received symbol, u_j, given the information about all parity check nodes except node i and $R_{i,j}$ to be the LLR that check node i is satisfied when u_j is fixed to 0 and 1 respectively. Given the vector $\boldsymbol{\Lambda}^{in}$ of initial LLRs, the BP algorithm outputs the extrinsic LLRs $\boldsymbol{\Lambda}^x$ as described below [78][52].

Algorithm 5.1. *Damped Log Belief Propagation (LBP)*
For all (i, j) such that $H_{i,j} = 1$:
Initialization: $Q_{i,j} = \Lambda_j^{in}$
DO

Horizontal Step:

$$R_{i,j} = \log\left(\frac{1 + \prod_{k \in J(i)\setminus j} \tanh(Q_{i,k}/2)}{1 - \prod_{k \in J(i)\setminus j} \tanh(Q_{i,k}/2)}\right)$$

$$= 2\tanh^{-1}\left(\prod_{k \in J(i)\setminus j} \tanh(Q_{i,k}/2)\right) \qquad (5.8)$$

Vertical Step:

$$Q_{i,j} = \Lambda_j^{in} + \theta \sum_{k \in I(j)\setminus i} R_{k,j}$$

While *stopping criterion is not met.*

Extrinsic Information: $\Lambda_j^i = \sum_{k \in I(j)} R_{k,j}.$

The factor θ is termed the *vertical step damping factor* and $0 < \theta \leq 1$. The magnitude of θ is determined by our level of confidence about the extrinsic information. In our implementations, θ is 0.5. Equation (5.8) is specifically useful for fast hardware implementations where the tanh function will be quantized to a reasonable accuracy and implemented as a lookup table. In our implementation, damped LBP is run for a small number of iterations on a fixed parity check matrix, so the stopping criterion is the number of iterations. In case that only one LBP iteration is run on the parity check matrix, the vertical step is eliminated.

Following we describe the Jiang-Narayanan algorithm [63, 64], which builds on the BP algorithm. In the JN algorithm, BP is run on the parity check matrix after reducing its independent columns corresponding to the least reliable bits to an identity submatrix. We will refer to such a class of algorithms, that adapt the parity check matrix before running BP, by adaptive belief propagation (ABP).

Algorithm 5.2. *The JN Algorithm*
Initialization: $\mathbf{\Lambda}^p := \mathbf{\Lambda}^{ch}$
DO

  1. Sort $\mathbf{\Lambda}^p$ in ascending order of magnitude and store the sorting index. The result-

ing vector of sorted LLRs is [1]

$$\mathbf{\Lambda}^{in} = [\Lambda_1^{in}, \Lambda_2^{in}, \ldots, \Lambda_{nm}^{in}],$$

$\|\Lambda_k^{in}\|_1 \leq \|\Lambda_{k+1}^{in}\|_1$ *for* $k = 1, 2, \ldots, nm - 1$ *and* $\mathbf{\Lambda}^{in} = P\mathbf{\Lambda}^p$ *where* P *defines a permutation matrix.*

2. *Rearrange the columns of the binary parity check matrix* H *to form a new matrix* H_P *where the rearrangement is defined by the permutation* P.

3. *Perform Gaussian elimination (GE) on the matrix* H_P *from left to right. GE will reduce the first independent* $(n - k)m$ *columns in* H_P *to an identity submatrix. The columns which are dependent on previously reduced columns will remain intact. Let this new matrix be* $\hat{H}_P$.

4. *Run log BP on the parity check matrix* $\hat{H}_P$ *with initial LLRs* $\mathbf{\Lambda}^{in}$ *for a maximum number of iterations* It_H *and a vertical step damping factor* θ. *The log BP algorithm outputs extrinsic LLRs* $\mathbf{\Lambda}^x$.

5. *Update the LLRs,* $\mathbf{\Lambda}^q = \mathbf{\Lambda}^{in} + \alpha_1\mathbf{\Lambda}^x$ *and* $\mathbf{\Lambda}^p := P^{-1}\mathbf{\Lambda}^q$ *where* $0 < \alpha_1 \leq 1$ *is called the ABP damping factor and* P^{-1} *is the inverse of* P.

6. *Decode using* $\mathbf{\Lambda}^p$ *as an input to the decoding algorithm* D.

While *Stopping criterion not satisfied.*

The JN algorithm assumed that the decoder D is one of the following hard-decision decoders:

- HD: Perform hard-decisions on the updated LLRs, $\hat{u} = (1 - \text{sign}(\mathbf{\Lambda}^p))/2$. If $H\hat{u}^T = 0$, then a decoding success is signaled.

- BM: Run a bounded minimum distance decoder such as the Berlekamp-Massey (BM) algorithm on the LLRs after hard-decisions. If the BM algorithm finds a codeword, a decoding success is signaled.

---

[1]To prevent notational ambiguity, $\|x\|_1$ will denote the magnitude of x.

The performance largely depends on the decoder D and the stopping criterion used. This is discussed in the following section.

5.3   Modifications to the Jiang-Narayanan Algorithm

The stopping criterion deployed in the JN algorithm is as follows [63]:

- Stop if a decoding success is signaled by the decoder D or if the number of iterations is equal to the maximum number of iterations, N_1.

We propose a list-decoding stopping criterion in which a list of codewords is iteratively generated. The list-decoding stopping criterion is as follows

- If a decoding success is signaled by the decoder D, add the decoded codeword to a *global* list of codewords. Stop if the number of iterations is equal to the maximum number of iterations, N_1.

If more than one codeword is on the global list of codewords, then the list decoder's output is the codeword which is at the minimum Euclidean distance from the received vector. Alternatively, one could only save the codeword with the largest conditional probability, given the received vector. This codeword would be the candidate for the list decoder's output when the iteration loop terminates.

The advantage of our proposed list-decoding stopping criterion over the stopping criterion in the JN algorithm is emphasized in the case of higher rate codes, where the decoder error probability is relatively high. Given a decoding algorithm D, the JN ABP algorithm may result in updating the received vector to lie in the decoding region of an erroneous codeword. However, running more iterations of the JN ABP algorithm may move the updated received vector into the decoding sphere of the transmitted codeword. The decoding algorithm D should also be run on the channel LLRs before any ABP iteration is carried out. If the decoder succeeds to find a codeword, it is added to the list.

Jiang and Narayanan [64] proposed running N_2 parallel decoders (outer iterations), each with the JN stopping criterion and a maximum of N_1 inner iterations. Each one of these N_2 iterations (decoders) starts with a different random permutation of the sorted channel LLRs in the first inner iteration. The outputs of these N_2 decoders form a list of at most N_2 codewords. If each of these N_2 decoders succeeds to find a codeword, the closest codeword to the received vector is chosen. We also run N_2 parallel decoders (outer iterations), each with the list-decoding stopping criterion, to form a global list of at most $N_1 N_2$ codewords. We propose doing the initial sorting of the channel LLRs in a systematic way to ensure that most bits will have a chance of being in the identity sub-matrix of the adapted parity check matrix. The improved performance achieved by these restarts could be explained by reasoning that if a higher reliability bit is in error, then it has a higher chance of being corrected if its corresponding column in the parity check matrix is in the sparse identity submatrix.

Let $z = \lfloor \tilde{n}/N_2 \rfloor$, then at the $(j+1)$st outer iteration, $j > 0$, the initial LLR vector at the first inner iteration is

$$[\Lambda_{jz+1}^{in}, \ldots, \Lambda_{(j+1)z}^{in}, \Lambda_1^{in}, \ldots, \Lambda_{jz}^{in}, \Lambda_{(j+1)z+1}^{in}, \ldots, \Lambda_{\tilde{n}}^{in}], \tag{5.9}$$

where $\mathbf{\Lambda}^{in}$ is the vector of sorted channel LLRs. The columns of H_P will also be rearranged according to the same permuatation. If $(j+1)z \leq \tilde{r}$, then it is less likely that this initial permutation will introduce new columns into the identity submatrix other than those which existed in the first outer iteration. After the first outer iteration, it is thus recommended to continue with the $(j+1)$st outer iteration such that $(j+1) > \tilde{r}/z$.

Another modification that could improve the performance of the JN algorithm is to run a small number of iterations of damped log belief propagation on the same parity check matrix. Although belief propagation is not exact due to the cycles in the associated Tanner graph [104], running a very small number of iterations of belief propagation is very effective [121]. Observing that the complexity of belief propagation is much lower than that of Gaussian elimination, one gets a performance enhancement at a slightly increased complexity.

Throughout the remaining of this chapter, we will refer to the modified JN algorithm with a list-decoding stopping criterion, as well as with the other modifications introduced in this section, by ABP-BM if the decoding algorithm D is BM (see Algorithm 5.2). Similarly, if the decoding algorithm was HD, the algorithm is referred to by ABP-HD. One of the main contribution in this chapter, the utilization of the a posteriori probabilities at the output of the ABP algorithm as the soft information input to an ASD algorithm, is presented in the following section.

5.4   The Hybrid ABP-ASD List Decoding Algorithm

Koetter and Vardy [72] point out that it is hard to maximize the mean of the *score* with respect to the to the true channel a posteriori probabilities. Previous multiplicity assignment algorithms [72, 83, 31] assumed approximate a posteriori probabilities. The problem is simplified by assuming that the transmitted codeword is drawn uniformly from $\mathbb{F}_q^n$. Also, the n received symbols are assumed to be independent and thus be assumed to be uniformly distributed. In such a case, the a posteriori probabilities are approximated to be a scaling of the channel transition probabilities,

$$\Pi_i^{ch}(\beta) = \frac{Pr\{y_i | u_i = \beta\}}{\sum_{\omega \in \mathbb{F}_q} Pr\{y_i | u_i = \omega\}}. \tag{5.10}$$

However, from the maximum distance separable (MDS) property of RS codes any k symbols (only) are k-wise independent and could be treated as information symbols and thus uniformly distributed. Thus these assumptions are more valid for higher rate codes and for memoryless channels. It is well known that belief propagation algorithms improve the reliability of the symbols by taking into account the geometry of the code and the correlation between symbols (see for example [78].) Due to the dense nature of the parity check matrix of the binary image of RS codes, running belief propagation directly will not result in a good performance. Because the Tanner graph associated with the parity check matrix of the binary image of RS codes has cycles, the marginals passed by the (log) belief propagation algorithm are no longer independent and the information starts to propagate in the loops.

Jiang and Narayanan [63] proposed a solution to this problem by adapting the parity check matrix after each iteration. When updating the check node reliability information $R_{i,j}$ (see (5.8)) corresponding to a pivot in a single weight column, the information $Q_{i,j}$ from any of the least reliable independent bits does not enter into the summation. One reason for the success of ABP is that the reliability information of the least reliable bits is updated by only passing the information from the more reliable bits to them. An analytical model for belief propagation on adaptive parity check matrices was recently proposed [3].

Our ABP-ASD algorithm is summarized by the following chain,

$$
\boldsymbol{u} \to \Pi^{ch} \xrightarrow{ABP} \hat{\Pi} \underbrace{\xrightarrow{\mathcal{A}} M \to}_{ASD} \hat{\boldsymbol{u}}, \tag{5.11}
$$

where $\boldsymbol{u}$ is the transmitted codeword, $\mathcal{A}$ is a multiplicity assignment algorithm, M is the multiplicity matrix and $\hat{\boldsymbol{u}}$ is the decoder output. In particular, the ABP-ASD list decoder is implemented by deploying the list decoder stopping criterion, proposed in the previous section, with an ASD decoding algorithm D (see Algorithm 5.2):

- ASD: Using $\boldsymbol{\Lambda}^p$ generate an $q \times n$ reliability matrix $\hat{\Pi}$ which is then used as an input to an multiplicity assignment algorithm to generate multiplicities according to the required interpolation cost. This multiplicity matrix is passed to the (modified) GS list-decoding algorithm. If the generated codeword list is not empty, the list of codewords is augmented to the *global* list of codewords. If only one codeword is required, the codeword with the highest reliability with respect to the channel LLR's $\boldsymbol{\Lambda}^{ch}$ is added to the global list.

In this chapter, the KV algorithm is used as the multiplicity assignment scheme. More efficient but more complex MA schemes could also be used [31]. The joint ABP-ASD algorithm corrects decoder failures (the received word does not lie in the decoding region centered around any codeword) of the ASD decoder D, by iteratively enhancing the reliability information of the received word, and thus moving the received word into the decoding region around a certain codeword. The decoding region in turn depends on the algorithm D and the designed interpolation cost. Furthermore, it attempts to eliminate decoder errors (the decoded codeword is not the transmitted codeword) by iteratively adding codewords to the global list of codewords and choosing the most probable one.

Since ASD is inherently a list-decoding algorithm with a larger decoding region, it is expected that ABP-ASD outperforms ABP-HD and ABP-BM. Since our algorithm transforms the channel LLRs into interpolation multiplicities for the GS algorithm, then, by definition, it is an interpolation multiplicity assignment algorithm for ASD.

The ABP-ASD algorithm has a polynomial-time complexity. The ABP step involves $o(\tilde{n}^2)$ floating point operations, for sorting and BP, and $o(\min(\tilde{k}^2, \tilde{r}^2)\,\tilde{n})$ binary operations for GE [64]. As for ASD, the KV MA algorithm (see (4.21)) has a time complexity of $O(n^2)$. An efficient algorithm for solving the interpolation problem is Koetter's algorithm [76] with a time complexity of $O(n^2\lambda^4)$. A reduced complexity interpolation algorithm is given in [80]. Roth and Ruckenstein [95] proposed an efficient factorization algorithm with a time complexity $O((l\log^2 l)k(n + l\log q))$, where l is an upper bound on the ASD's list size and is determined by λ.

5.5   A Low Complexity ABP Algorithm

Most of the complexity of adaptive belief propagation lies in row reducing the binary parity check matrix (after rearranging the columns according to the permutation P). To reduce the complexity one could make use of the columns already reduced in the previous iteration.

We will use the same notation as in Algorithm 5.2 with a subscript j to denote the values at iteration j. For example, the vector of sorted LLRs at the jth iteration is $\mathbf{\Lambda}_j^{in}$. Define $P_j(H)$ to be the matrix obtained when the columns of the parity check matrix H are permuted according to the permutation P_j at the jth iteration. $GE(H)$ will be the reduced matrix (with an identity submatrix) after Gaussian elimination is carried out on the matrix H.

Let $R_j \overset{\Delta}{=} \{t\ :\ t$th column of H was reduced to a column of unit weight in $GE(P_j(H))\}$. It is clear that the cardinality of R_j is $\tilde{r}$. Now assume that log BP is run and that the LLRs are updated and inverse permuted to get $\mathbf{\Lambda}_j^p$ (step 5 in Algorithm 5.2). The set of indices of the $\tilde{r}$ (independent) LLRs in $\mathbf{\Lambda}_j^p$ with the smallest magnitude will be denoted by S_{j+1}. By definition, P_{j+1} is the permutation that sorts the LLRs in $\mathbf{\Lambda}_j^p$ in ascending order according to their magnitude to get $\mathbf{\Lambda}_{j+1}^{in}$. The set $U_{j+1} \triangleq R_j \cap S_{j+1}$ is thus the set of indices of bits which are among the least reliable independent bits at the $(j + 1)$st iteration and whose corresponding columns in the reduced parity check matrix at the previous iteration were in the identity submatrix.

The algorithm is modified such that GE will be run on the matrix whose left most columns are those corresponding to U_{j+1}. To construct the identity submatrix, these columns may only require row permutations for arranging the pivots (ones) on the diagonal. Note that these permutations may have also been required when running GE on $P_{j+1}(H)$. Only a small fraction of the columns will need to be reduced to unit weight leading to a large reduction in the GE computational complexity. Also note that what matters is that a column corresponding to a bit with low reliability lies in the identity (sparse) submatrix and not its position within the submatrix. This is justified by the fact that the update rules for all the LLRs corresponding to columns in the identity submatrix are the same. Thus provided that the first $\tilde{r}$ columns in $P_{j+1}(H)$ are independent, changing their order does not alter the performance of the ABP algorithm. To summarize the proposed reduced complexity ABP algorithm can be stated as follows:

Algorithm 5.3. *Low Complexity Adaptive Belief Propagation*
Initialization: $\mathbf{\Lambda}^p := \mathbf{\Lambda}^{ch}, j = 1$
DO
If $j = 1$
Proceed as in the first iteration of Algorithm 5.2; $\mathbf{\Lambda}_1^{in} = \mathbf{\Lambda}^{in}|_{Algorithm\ 5.2}$, $P_1 = P|_{Algorithm\ 5.2}$, $\hat{H}_1 = \hat{H}_P|_{Algorithm\ 5.2}$ *and* $\mathbf{\Lambda}_1^q = \mathbf{\Lambda}^q|_{Algorithm\ 5.2}$.
If $j > 1$

1. *Sort the updated LLR vector $\mathbf{\Lambda}_{j-1}^q$ in ascending order of the magnitude of its elements. Let W_j' be the associated sorting permutation matrix.*

2. *Rearrange the columns of the binary parity check matrix $\hat{H}_{j-1}$ to form a new matrix*
$$Q_j' = W_j'(\hat{H}_{j-1}).$$

3. *Rearrange the left-most $\tilde{r}$ columns of the binary parity check matrix Q'_j such that the columns of unit weight are the most left columns. Let W''_j be the corresponding permutation matrix. (This could be done by sorting the first $\tilde{r}$ columns of Q'_j in ascending order according to their weight.) Let the resulting matrix be*

$$Q''_j = W''_j(Q'_j).$$

4. *Permute the LLR vector;*
$$\Lambda^{in}_j = P'_j \Lambda^q_{j-1},$$

where $P'_j = W'_j W''_j$.

5. *Update the (global) permutation matrix;*

$$P_j = P'_j P_{j-1}.$$

6. *Run Gaussian elimination on the matrix Q''_j from left to right;*

$$\hat{H}_j = GE(Q''_j).$$

7. *Run damped LBP on $\hat{H}_j$ with initial LLRs Λ^{in}_j for It_H iterations. The output vector of extrinsic LLRs is Λ^x_j.*

8. *Update the LLRs;*

$$\Lambda^q_j = \Lambda^{in}_j + \alpha_1 \Lambda^x_j \text{ and } \Lambda^p_j = P^{-1}_j \Lambda^q_j.$$

9. *Decode using Λ^p_j as an the input to the decoding algorithm D.*

10. *Increment j.*

While *Stopping criterion not satisfied.*

The algorithm as described above iteratively updates a global permutation matrix and avoids inverse permuting the row-reduced parity check matrix in each iteration.

The implementation of the algorithm also assumes for simplicity that the columns in the parity check matrix corresponding to the $\tilde{r}$ least reliable bits are independent and could therefore be reduced to unit weight columns. It is also noticed that in practice the cardinality of U_{j+1} is close to $\tilde{r}$ which means that the GE elimination complexity will be significant only in the first iteration.

We will assume the favorable condition in which the most left $\tilde{r}$ columns of an parity check matrix are independent. Taking into account that the parity check matrix is a binary matrix, the maximum number of binary operations required to reduce the first $\tilde{r}$ columns to an identity submatrix in the JN algorithm (Algorithm 5.2) can be shown to be

$$\Theta_{GE} = 2\sum_{\alpha=1}^{\tilde{r}}(\tilde{r} - \alpha)(\tilde{n} - \alpha + 1) < \tilde{r}^2\tilde{n} - \tilde{r}\tilde{k}. \tag{5.12}$$

(It is assumed that the two GE steps, elimination and back substitution, are symmetric.) Row permutation operations were neglected. Now assume that the cardinality of U_{j+1} is $\delta\tilde{r}$, where $\delta \leq 1$.

For the modified algorithm, only row permutations may be required for the first $\delta\tilde{r}$ columns to arrange the pivots on the diagonal of the identity submatrix. These permutations may also be required for the JN algorithm. Then the relative reduction in complexity is

$$\frac{\Theta_{GE} \text{ in Algorithm } 5.2 - \Theta_{GE} \text{ in Algorithm } 5.3}{\Theta_{GE} \text{ in Algorithm } 5.2} =$$
$$\frac{\sum_{\alpha=1}^{\delta\tilde{r}}(\tilde{r} - \alpha)(\tilde{n} - \alpha + 1)}{\sum_{\alpha=1}^{\tilde{r}}(\tilde{r} - \alpha)(\tilde{n} - \alpha + 1)} \approx$$
$$\frac{(\tilde{r}^2\tilde{n})(2\delta - \delta^2) - \delta\tilde{r}\tilde{k}}{\tilde{r}^2\tilde{n} - \tilde{r}\tilde{k}} \approx 2\delta - \delta^2. \tag{5.13}$$

For example, if we assume that on average $\delta = 0.5$, a simple calculation for the $(255, 239)$ code over F_{256} shows that the relative reduction in the complexity of the GE step is about 75%. In practice δ is close to one. Note that Algorithm 5.3 does require sorting $\tilde{r}$ columns of Q'_j according to their weight (step 3) but the complexity is relatively small.

5.6   Numerical Results and Discussion

In the next subsection, a fast simulation setup is described for ABP list decoding. Bounds on the error probability of the ML decoder are then discussed. We then show simulation results for our algorithm.

5.6.1   Fast Simulation Setup

We describe a fast simulation setup for ABP with a list-decoding stopping criterion. One could avoid running the actual decoder D at each iteration and instead check whether the transmitted codeword is on the list generated by the decoder D. The stopping criterion would be modified such that the iterative decoding stops if the transmitted codeword is on the list or if the maximum number of iterations is reached. A decoding success is signaled if the transmitted codeword is on the list.

It is easy to see that this simulation setup is equivalent to running the actual ABP list decoder for the maximum number of iterations. Suppose that the received sequence results in an maximum-likelihood (ML) error, then it is very unlikely that the decoder D will correctly decode the received word at any iteration. In case of an ML decoder success and the transmitted codeword is added to the global list at a certain iteration, which presumably could be checked, then it would be the closest codeword to the received word and thus the list decoder's choice. Thus for a fast implementation, a decoding success is signaled and iteration stops once the transmitted codeword appears on the global list.

In case that D is a bounded minimum distance decoder such as the Berlekamp-Massey (BM) algorithm, the transmitted codeword would be on the global list if it is at a Hamming distance of $\lfloor \frac{n-k}{2} \rfloor$ or less from the hard-decisioned (modified) LLRs. If D is an ASD algorithm that assigns the multiplicity matrix M, the transmitted codeword is on the ASD's list (and thus the global list) if it satisfies the sufficient conditions of (4.4) and (4.6) for finite and infinite interpolation costs respectively. It was shown in [72], that simulating the KV algorithm by checking the sufficient condition of (4.4) results in accurate results. This is partially justified by the fact that on average, the ASD's list size is one [77]. This is also justified by observing that if the ASD's list is

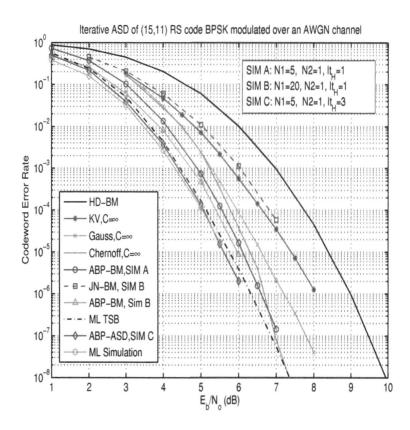

Figure 5.1: The performance of iterative ASD of $(15, 11)$ RS code, BPSK modulated over an AWGN channel, is compared to that of other ASD algorithms and ABP-BM list decoding.

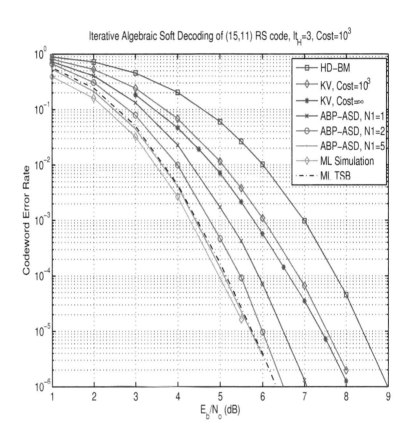

Figure 5.2: The performance of iterative ASD of the $(15, 11)$ RS code, BPSK modulated over an AWGN channel, is shown for a finite interpolation cost of 10^3 and different iteration numbers.

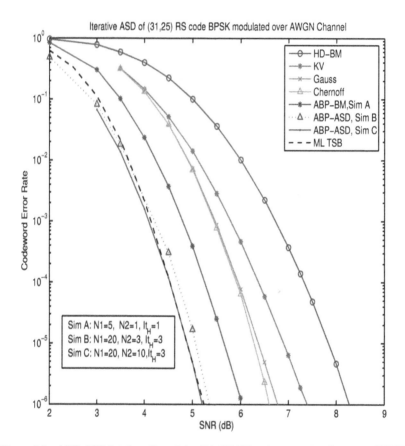

Figure 5.3: ABP-ASD list decoding of the $(31, 25)$ RS code transmitted over an AWGN with BPSK modulation.

empty (a decoding failure), the condition (4.4) will not be satisfied. However, if the list is nonempty but the transmitted codeword is not on the list (a decoding error), the condition will still not be satisfied for the transmitted codeword and a decoding error/failure is signaled. However if the condition is satisfied, then this implies that the transmitted codeword is on the ASD's list and thus a decoding success.

5.6.2   Bounds on the Maximum-Likelihood Error Probability

As important as it is to compare our algorithms with other algorithms, it is even more important to compare it with the ultimate performance limits, which is that of the soft-decision ML decoder. When transmitting the binary image of RS codes over a channel, the performance of the maximum-likelihood decoder depends on the weight enumerator of the transmitted binary image. The binary image of RS codes is not unique, but depends on the basis used to represent the symbols as bits. An average binary weight enumerator of RS codes could be derived by assuming a binomial distribution of the bits in a nonzero symbol [29]. Based on the Poltyrev tangential sphere bound (TSB) [87] and the average binary weight enumerator, average bounds on the ML error probability of RS codes over additive white Gaussian noise (AWGN) channels were developed in [29] and were shown to be tight. We will refer to this bound by ML-TSB. Alternatively the averaged binary weight enumerator could be used in conjunction with other tight bounds such as the Divsalar simple bound [23] to bound the ML error probability. We refer the reader to Chapter 2 for more information.

5.6.3   Numerical Results

In this subsection, we give some simulation results for our algorithm. As noted before, the multiplicity assignment algorithm used for ABP-ASD in the these simulations is the KV algorithm. $N2$ denotes the number of outer iterations (parallel decoders) and $N1$ is the number of inner iterations in each of these outer iterations.

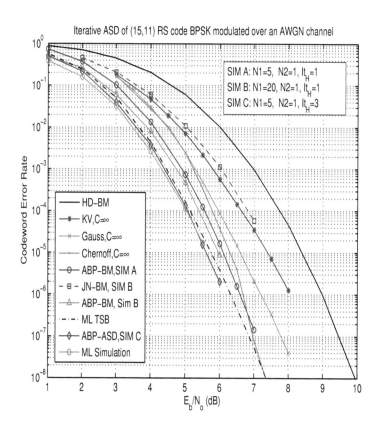

Figure 5.4: The performance of iterative ASD of $(15, 11)$ RS code, BPSK modulated over an AWGN channel, is compared to that of other ASD algorithms and ABP-BM list decoding.

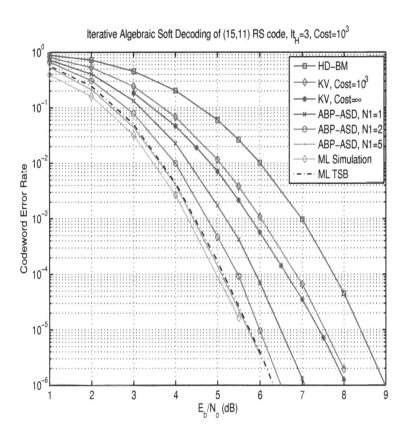

Figure 5.5: The performance of iterative ASD of the $(15, 11)$ RS code, BPSK modulated over an AWGN channel, is shown for a finite interpolation cost of 10^3 and different iteration numbers.

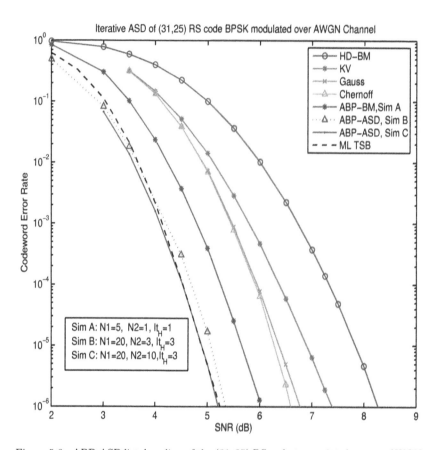

Figure 5.6: ABP-ASD list decoding of the $(31, 25)$ RS code transmitted over an AWGN with BPSK modulation.

$(15, 11)$ RS Code over an AWGN Channel

A standard binary input AWGN channel is assumed where the transmitted codewords are BPSK modulated. In Figure 5.4, we compare the performance of different decoding algorithms. HD-BM refers to the performance of a hard decision bounded minimum distance decoder such as the BM algorithm. The ABP-BM list-decoding algorithm with $N1 = 5$ iterations and one iteration of LBP on each parity check matrix, $It_H = 1$ (see step 4 in Algorithm 5.2) has a coding gain of about 2.5 dB over HD-BM at a codeword error rate (CER) of 10^{-6}. Increasing the number of iterations to $N1 = 20$ iterations, we get a slightly better performance. JN-BM refers to the JN algorithm with the JN stopping criterion and a BM decoder. Due to the high decoder error probability of the $(15, 11)$ code, ABP-BM, with the list decoder stopping criterion, yields a much better performance than JN-BM. The ABP-ASD list-decoding algorithm outperforms all the previous algorithms with only 5 ABP iterations and with $It_H = 3$. Comparing its performance with soft-decision ML decoding of the RS code, we see that ABP-ASD has a near ML performance with a performance gain of about 3 dB over HD-BM at a CER of 10^{-6}. (ML decoding was carried out by running the BCJR algorithm on the trellis associated with the binary parity check matrix of the RS code [66].) Moreover, the averaged TSB on the ML codeword error probability is shown to confirm that it is a tight upper bound and that the ABP-ASD algorithm is near optimal for this code.

The performance of different ASD algorithms are compared for infinite interpolation costs, the KV algorithm [72], the Gaussian approximation (Gauss) [83] and the Chernoff bound algorithm (Chernoff) [31]. It is noted that the Chernoff bound algorithm has the best performance, especially at the tail of error probability. It is also interesting to compare the performance of ABP-ASD with other ASD MA algorithms. It has about 2 dB coding gain over the KV algorithm at a CER of 10^{-6}. As expected, the Chernoff method has a comparable performance at the tail of the error probability.

The ABP algorithm used in the simulations shown in Figure 5.4 is Algorithm 5.2. The performance of Algorithm 5.3 was identical to that of Algorithm 5.2. However, the complexity is much less. The average δ (see (5.13)) averaged over all iterations was calculated versus the SNR. It was observed that the ratio of the number of columns to be reduced in Algorithm 5.3 to that in Algorithm 5.2 is about 0.1 ($\delta = 0.9$). This gives

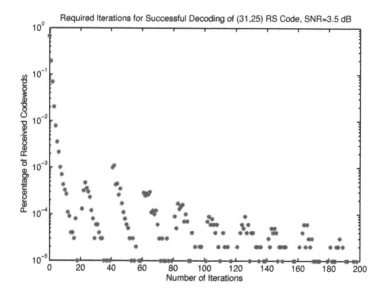

Figure 5.7: Convergence of the iterative ASD algorithm.
This histogram shows the percentage of transmitted codewords successfully decoded versus the iteration number at which the transmitted codeword was first successfully added to the ABP-ASD list with $N1 = 20$ and $N2 = 10$. The $(31, 25)$ RS code is transmitted over an AWGN channel at an SNR of 3.5 dB.

about a 99% reduction in the Gaussian elimination complexity. Thus only the first iteration or restart suffers from an Gaussian elimination complexity if Algorithm 5.3 is used.

Near ML decoding for the same code is also achieved by the ABP-ASD algorithm with a finite cost of 10^3 as shown in Figure 5.5. Comparisons are made between the possible coding gains if the number of iterations is limited to $N1 = 1, 2, 5$. With 5 iterations, the performance gain over the KV algorithm, with the same interpolation cost, is nearly 1.8 dB at a CER of 10^{-5}. Comparing the ABP-ASD performance to that of Figure 5.4, with infinite interpolation costs, we observe that a small loss in performance results with reasonable finite interpolation costs. Unless otherwise stated, the remaining simulations in this chapter will assume infinite interpolation costs to show the potential of our algorithm.

It is to be noted that in simulating the ABP-BM list decoder, the simulations using a real BM decoder were identical to the simulations using the fast simulation setup described in this section. To save simulation time, the curves shown here for ABP-ASD are generated using the fast simulation setup. As is the case for ABP-BM, running the real ABP-ASD decoder will yield the same results.

$(31, 25)$ RS Code over AWGN Channel

The arguments for the $(15, 11)$ RS code also carry over for the $(31, 25)$ RS code when BPSK modulated and transmitted over an AWGN channel, as shown in Figure 5.6. With only 5 iterations, the ABP-BM list-decoding algorithm outperforms previous ASD algorithms. The performance of ABP-ASD with 20 inner iterations (N1) and 10 outer iterations (N2) is better than the ML upper bound and has more than 3 dB coding gain over the BM algorithm at an CER of 10^{-4}. A favorable performance is also obtained by only 3 restarts (outer iterations). By comparing with Figure 5.5 of [103], our ABP-ASD algorithm has about 1.6 dB gain over the combined Chase II-GMD algorithm at an CER of 10^{-4}.

To show the effectiveness of the restarts or outer iterations, we kept track of the iteration number at which the ABP-ASD list decoder was first capable to successfully decode the received word. In other words, this is the iteration when the transmitted

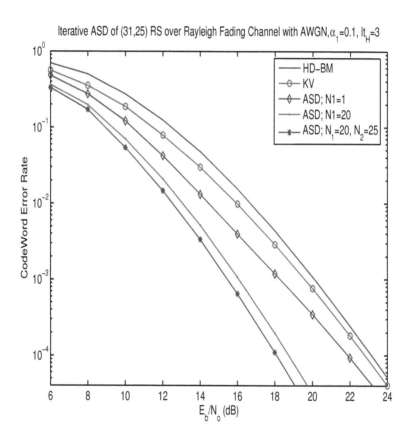

Figure 5.8: The performance of the ABP-ASD decoding of the $(31, 25)$ RS code over a Rayleigh fading channel with AWGN when the channel information is unknown at the decoder.

codeword was first added to the ABP-ASD list. The percentage of transmitted codewords which were first successfully decoded at a certain iteration is plotted versus the iteration number in the histogram of Figure 5.7. This is shown at a signal-to-noise ratio (SNR) of 3.5 dB and for $N1 = 20$ $N2 = 10$ with a total of 200 iterations. At the beginning of each restart (every 20 iterations) there is a boost in the number of codewords successfully decoded and this number declines again with increasing iterations. The zeroth iteration corresponds to the KV algorithm. This histogram is also invaluable for decoder design and could aid one to determine the designed number of iterations for a required CER.

$(31, 25)$ RS Code over a Rayleigh Fading Channel

As expected from the discussion in Section 5.4, the coding gain of ABP-ASD is much more if the underlying channel model is not memoryless. This is demonstrated in Figure 5.8, where an $(31, 25)$ code is BPSK modulated over a relatively fast Rayleigh fading channel with AWGN. The Doppler frequency is equal to 50 Hz and the codeword duration is 0.02 seconds. The coding gain of ABP-ASD over the KV algorithm at an CER of 10^{-4} is nearly 5 dB when the channel is unknown to both decoders.

$(255, 239)$ RS Code over AWGN Channel

The performance of the ABP-ASD algorithm is also investigated for relatively long codes. The $(255, 239)$ code and its shortened version, the $(204, 188)$ code, are standards in many communication systems. The performance of the $(255, 239)$ code over an AWGN channel is shown in Figure 5.9. By 20 iterations of ABP-BM, one could achieve a coding gain of about 0.5 dB over the KV algorithm. At an CER of 10^{-6}, after a total of 25 outer iterations (restarts), the coding gain of ABP-ASD over BM is about 1.5 dB. An extra 0.1 dB of coding gain is obtained with 25 more outer iterations. Moreover, the performance of the ABP-ASD decoder is within 1 dB of the averaged ML TSB.

$(31, 15)$ RS Code over AWGN Channel

The performance of our algorithm is studied for the $(31, 15)$ RS code over an AWGN channel. The rate of this code is 0.48. Because this code is of relatively low rate, the

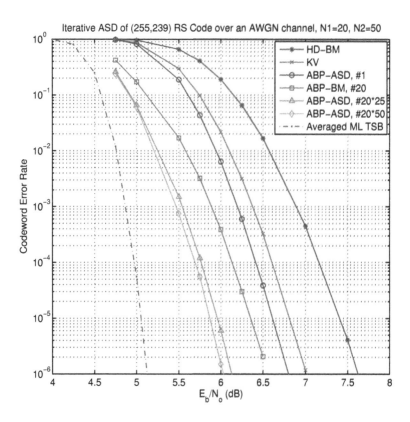

Figure 5.9: The performance of the ABP-ASD decoding of the $(255, 239)$ RS code over an AWGN channel with BPSK modulation.

HD-GS algorithm does improve over the HD-BM bounded minimum distance decoding algorithm. As seen from Figure 5.10, ML soft-decision decoding offers about 4 dB coding gain over the hard decision GS algorithm and about 2.8 dB coding gain over the soft-decision KV ASD algorithm at an CER of 10^{-5}. With 20 iterations, ABP-BM list decoding improves over the KV algorithm. As expected, ABP-ASD has a better performance for the same number of iterations. With 10 restarts, ABP-ASD has a reasonable performance with about a 3 dB coding gain over the BM algorithm. Another 0.5 dB of coding gain could be achieved by increasing the number of iterations.

General Observations

It is noticed that the coding gain between iterations decreases with the number of iterations. It is also to be noted that the ABP-ASD list decoder requires running the KV ASD algorithm in each iteration. Running a number of 'plain-vanilla' ABP iterations without the ASD decoder and then decoding using the ASD decoder (to reduce the complexity) will yield a worse performance for the same number of iterations. The same arguments also hold for the ABP-BM list decoding. A reasonable performance is achieved by ABP-BM list decoding. By deploying the KV ASD algorithm, ABP-ASD list decoding has significant coding gains over the KV ASD algorithm and other well-known soft-decision decoding algorithms.

5.7   Conclusion

In this chapter, we proposed a list-decoding algorithm for soft-decision decoding of Reed-Solomon codes. Our algorithm is based on enhancing the soft reliability channel information before passing them to an algebraic soft-decision decoding algorithm. This was achieved by deploying the Jiang and Narayanan algorithm, which runs belief propagation on an adapted parity check matrix. Using the Koetter-Vardy algorithm as the algebraic soft-decision decoding algorithm, our algorithm has impressive coding gains over previously known soft-decision decoding algorithms for RS codes. By comparing with averaged bounds on the performance of maximum-likelihood decoding of RS codes, we observe that our algorithm achieves a near optimal performance for rela-

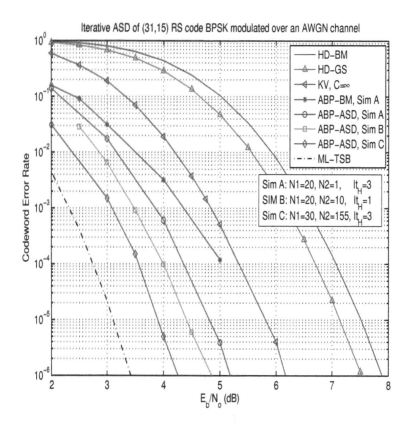

Figure 5.10: ABP-ASD list decoding of the $(31, 15)$ RS code, of rate 0.48, transmitted over an AWGN with BPSK modulation.

tively short, high-rate codes. We introduced some modifications over the JN algorithm that resulted in better coding gains. We presented a low complexity adaptive belief-propagation algorithm, which results in a significant reduction in the computational complexity. The performance of our algorithm was studied for the cases when the interpolation cost of the algebraic soft-decision decoding algorithm is both finite and infinite. A small loss in coding gain results when using manageable interpolation costs. The coding gain of the presented algorithm is larger for channels with memory. Our proposed algorithm could also be viewed as an interpolation multiplicity assignment algorithm for algebraic soft decoding.

The question remains whether the JN algorithm is the optimum way to process the channel reliability information before algebraic soft-decision decoding. The KV algorithm was our ASD decoder of choice due to its low complexity. It would be interesting to determine the best ASD algorithm or, in general, soft-decision decoding algorithm for joint belief propagation. From another point of view, suppose we have the KV ASD algorithm, it is also interesting to determine the best low complexity technique that will process the channel reliability information before passing it to the ASD algorithm. The tradeoff between performance and computational complexity is likely to play a big role in determining the state-of-art next generation Reed-Solomon decoders.

Chapter 6

Performance Analysis of Linear Product Codes

Big doors swing on little hinges.

—W. Clement Stone

Product codes were introduced by Elias [38] in 1954, who also proposed to decode them by iteratively (hard) decoding the component codes. With the invention of turbo codes [13], soft iterative decoding techniques received wide attention [52]: low complexity algorithms for turbo decoding of product codes were first introduced by Pyndiah in [90]. Other efficient algorithms were recently proposed in [57] and in [6].

For product codes, an interesting issue for both theory and applications regards the analytical estimation of their maximum-likelihood performance. Among other, this analytical approach allows one to (i) forecast the code performance without resorting to simulation, (ii) provide a benchmark for testing suboptimal iterative decoding algorithms, (iii) establish the goodness of the code, determined by the distance from theoretical limits.

The analytical performance evaluation of a maximum-likelihood decoder requires the knowledge of the code weight enumerator. Unfortunately, the complete enumerator is unknown for most families of product codes. In these years, some progress has been made in determining the first terms of product code weight enumerators. The multiplicity of low weight codewords for an arbitrary linear product code were computed by Tolhuizen [105]. (In this chapter, these results will be extended to find the exact input-output weight enumerators of low weight codewords.)

Even if the first terms can be individuated, the exact determination of the complete weight enumerator is very hard for arbitrary product codes [105], [26]. By approximating the number of the remaining codewords by that of a normalized random code, upper bounds on the performance of binary product codes using the ubiquitous union bound were shown in [106]. However, this approximation is not valid for all product codes.

In this chapter, we will consider the representation of a product codes as a concatenated scheme with interleaver, and we will derive the average input-output weight enumerator for linear product codes over a generic field $\mathbb{F}_q$. When combined with the extended Tolhuizen's result, this will provide a complete approximated enumerator for the product code. We will show how it closely approximates the exact weight enumerator.

Previous work in the literature (see for example [19], and reference therein) focused on estimating the product code performance at low error rates via the truncated union bound, using the enumerator low-weight terms only. By using the complete approximate enumerator, it is possible to compute the Poltyrev bound [87], which establish tight bounds on the maximum-likelihood performance at both high and low error rates.

The outline of the chapter is as follows. In Section 6.1, we introduce the basic notation and definitions. In Section 6.2, we extend Tolhuizen results and derive the exact input-output weight enumerator for product code low-weight codewords. Product code representation as serial and parallel concatenated codes with interleavers are introduced in Section 6.3.1. Uniform interleavers on finite fields with arbitrary size are discussed in Section 6.3.2. The average weight enumerators of product codes are then computed in Section 6.3.3. The merge with exact low-weight terms, and the discussion of the combined enumerator properties are performed in Section 6.4.

The computation of product code average enumerators relies on the knowledge of the input-redundancy weight enumerators of the component codes. For this reason, we derive in Section 6.5 closed form formulas for the enumerator functions of some linear codes commonly used in the construction of product codes: Hamming, extended

Hamming, and Reed-Solomon codes. We proceed in Section 6.6 to derive the average binary weight enumerators of Reed-Solomon product codes defined on finite fields of characteristic two.

To support our theory, we present some numerical results. Complete average enumerators are depicted and discussed in Section 6.7.1. Analytical bounds on the maximum-likelihood performance are shown at both high and low error rates, and compared against simulation results in Section 6.7.2. Finally, we conclude the chapter in Section 6.8.

6.1   Preliminaries

As in the previous chapters, $\mathbb{F}_q$ will be a finite field of q elements. $\mathcal{C}$ will denote an (n_c, k_c, d_c) linear code over $\mathbb{F}_q$ with codeword length n_c, information vector length k_c and minimum Hamming distance d_c. Let $\mathcal{R}$ and $\mathcal{C}$ be (n_r, k_r, d_r) and (n_c, k_c, d_c) linear codes over $\mathbb{F}_q$, respectively. The product code whose component codes are $\mathcal{R}$ and $\mathcal{C}$, $\mathcal{P} \triangleq \mathcal{R} \times \mathcal{C}$, consists of all matrices such that each row is a codeword in $\mathcal{R}$ and each column is a codeword in $\mathcal{C}$. $\mathcal{P}$ is an (n_p, k_p, d_p) linear code, with parameters

$$n_p = n_r n_c \qquad k_p = k_r k_c \qquad d_p = d_r d_c.$$

We will recall some definitions and results from Chapters 2 and Chapters 3. The *weight enumerator* (WE) of $\mathcal{C}$, $E_{\mathcal{C}}(h)$, is the number of codewords with Hamming weight h:

$$E_{\mathcal{C}}(h) = |\{\boldsymbol{c} \in \mathcal{C} : \mathcal{W}(\boldsymbol{c}) = h\}|,$$

where $\mathcal{W}(\cdot)$ denotes the symbol Hamming weight. For a systematic code $\mathcal{C}$, the *input-redundancy weight enumerator* (IRWE), $R_{\mathcal{C}}(w, p)$, is the number of codewords with information vector weight w, whose redundancy has weight p:

$$R_{\mathcal{C}}(w, p) = |\{\boldsymbol{c} = (\boldsymbol{i}|\boldsymbol{p}) \in \mathcal{C} : \mathcal{W}(\boldsymbol{i}) = w \quad \mathcal{W}(\boldsymbol{p}) = p\}|.$$

If $\mathcal{T} = (n_1, n_2)$ is a partition of the n coordinates of the code into two sets of size n_1

and n_2, the *split weight enumerator* $A^T(w_1, w_2)$ is number of codewords with Hamming weights w_1 and w_2 in the first and second partition, respectively. If T is an $(k, n-k)$ partition such the first set of cardinality k constitutes of the information symbol coordinates, $R(w_1, w_2) = A^T(w_1, w_2)$. The *input-output weight enumerator* (IOWE), $O_{\mathcal{C}}(w, h)$, is the number of codewords whose Hamming weight is h, while their information vector has Hamming weight w:

$$O_{\mathcal{C}}(w, h) = |\{\boldsymbol{c} \in \mathcal{C} : \mathcal{W}(\boldsymbol{i}) = w \quad \mathcal{W}(\boldsymbol{c}) = h\}|.$$

For a systematic code,

$$O_{\mathcal{C}}(w, h) = R_{\mathcal{C}}(w, h - w). \tag{6.1}$$

It is also straightforward that

$$E_{\mathcal{C}}(h) = \sum_{w=0}^{k_c} O_{\mathcal{C}}(w, h). \tag{6.2}$$

The WE *generating function* of $\mathcal{C}$ is defined by this polynomial in invariant Y:

$$\mathbb{E}_{\mathcal{C}}(Y) = \sum_{h=0}^{n_c} E_{\mathcal{C}}(h) Y^h$$

while the IRWE function and the IOWE function of $\mathcal{C}$ are defined by these bivariate polynomials in invariants X and Y:

$$\mathbb{R}_{\mathcal{C}}(X, Y) = \sum_{w=0}^{k_c} \sum_{p=0}^{n_c - k_c} R_{\mathcal{C}}(w, p) X^w Y^p, \tag{6.3}$$

$$\mathbb{O}_{\mathcal{C}}(X, Y) = \sum_{w=0}^{k_c} \sum_{h=0}^{n_c} O_{\mathcal{C}}(w, h) X^w Y^h. \tag{6.4}$$

These functions are related by

$$\mathbb{O}_{\mathcal{C}}(X, Y) = \mathbb{R}_{\mathcal{C}}(XY, Y), \tag{6.5}$$

$$\mathbb{E}_{\mathcal{C}}(Y) = \mathbb{R}_{\mathcal{C}}(Y, Y) = \mathbb{O}_{\mathcal{C}}(1, Y). \tag{6.6}$$

As in (2.3), we will denote the coefficient of $X^w Y^h$ in a bivariate polynomial $\mathbb{Q}(X, Y)$ by the coefficient function $\mathrm{Coeff}(\mathbb{Q}(X, Y), X^w Y^h)$. For example,

$$O_{\mathcal{C}}(w, h) = \mathrm{Coeff}(\mathbb{O}_{\mathcal{C}}(X, Y), X^w Y^h).$$

Similarly, $\mathrm{Coeff}(\mathbb{O}(X, Y), Y^w)$ is the coefficient of Y^w in the bivariate polynomial $\mathbb{O}(X, Y)$ and is a univariate polynomial in X.

Let the code $\mathcal{C}$ be transmitted by a binary PSK constellation over an AWGN channel with a signal-to-noise ratio (SNR) γ. As in Section 3.5, the the codeword error probability (CEP) and bit error probability (BEP) of the decoder will be denoted by $\Phi_c(E_{\mathcal{C}}(h), \gamma)$ and $\Phi_b(\gamma)$.

The truncated union bound, taking into account the minimum distance term only, provides a heuristic lower bound on the performance of soft-decision maximum-likelihood decoder:

$$\Phi_c(\gamma) \gtrsim \frac{1}{2} E_{\mathcal{C}}(d_c) \; \mathtt{erfc} \; \sqrt{\frac{k_c}{n_c} d_c \gamma} \; . \tag{6.7}$$

This formula provides a simple way for predicting the code performance at very high SNR/low CEP, where maximum-likelihood error events are mostly due to received noisy vectors lying in the decoding regions of codewords nearest to the transmitted one. Anyway, it is not useful in predicting the performance at low SNR.

Tight bounds on the maximum-likelihood codeword error probability of binary linear codes for AWGN and binary symmetric channel (BSC), holding at both low and high SNR, were derived by Poltyrev in [87]. These bounds usually require knowledge of the complete weight enumerator $E_{\mathcal{C}}(h)$ (c.f., Section 2.4). In this chapter, we will apply the Poltyrev bounds by using a complete approximate weight enumerator of the considered product codes.

Given the codeword error probability, the computation of the bit error probability may pose a number of technical problems. Let $\Phi_c(E_{\mathcal{C}}(h), \gamma)$ denote the CEP over a channel with an SNR γ computed by using the weight enumerator $E_{\mathcal{C}}(h)$. The bit error probability $\Phi_b(\gamma)$ is derived from the CEP by computing $\Phi_b(\gamma) = \Phi_c(I_{\mathcal{C}}(h), \gamma)$, where $I_{\mathcal{C}}(h) = \sum_{w=1}^{k_c} \frac{w}{k_c} O(w, h)$ (c.f., Section 3.5). A common approximation in the literature is $I_{\mathcal{C}}(h) \approx \frac{h}{n_c} E_{\mathcal{C}}(h)$. This approximation is useful if the IOWE $O(\cdot, \cdot)$ is not known but the weight enumerator WE $E(\cdot)$ is. Some codes satisfy this approximation with equality: they are said to possess the *multiplicity property*. We refer the reader to Section 3.3 for a discussion on such codes.

6.2 Exact IOWE of Low-Weight Codewords

Tolhuizen showed that in a linear product code $\mathcal{P} = \mathcal{R} \times \mathcal{C}$ the number of codewords with symbol Hamming weight $1 \leq h < h_o$ is [105]:

$$E_{\mathcal{P}}(h) = \frac{1}{q-1} \sum_{i|h} E_{\mathcal{C}}(i) E_{\mathcal{R}}(h/i), \qquad (6.8)$$

where, given

$$w(d_r, d_c) \overset{\Delta}{=} d_r d_c + \max(d_r \lceil \frac{d_c}{q} \rceil, d_c \lceil \frac{d_r}{q} \rceil),$$

the weight h_o is

$$h_o = \begin{cases} w(d_r, d_c) + 1, & \text{if } q = 2 \text{ and both } d_r \text{ and } d_c \text{ are odd} \\ w(d_r, d_c), & \text{otherwise} \end{cases} . \qquad (6.9)$$

In particular, the minimum distance multiplicity of a product code is given by

$$E_{\mathcal{P}}(d_p) = \frac{E_{\mathcal{R}}(d_r) E_{\mathcal{C}}(d_c)}{q - 1}. \qquad (6.10)$$

These results are based on the properties of *obvious* (or *rank-one*) codewords of $\mathcal{P}$, i.e., direct product of a row and a column codeword [105]. Let $\mathbf{r} \in \mathcal{R}$ and $\mathbf{c} \in \mathcal{C}$, then an obvious codeword, $\mathbf{p} \in \mathcal{P}$, is defined as

$$\mathbf{p}_{i,j} = \mathbf{r}_i \mathbf{c}_j, \qquad (6.11)$$

where $\mathbf{r}_i$ is the symbol in the ith coordinate of $\mathbf{r}$ and $\mathbf{c}_j$ is the symbol in the jth coordinate of $\mathbf{c}$. It follows that the rank of the $n_c \times n_r$ matrix defined by $\mathbf{p}$ is one and the Hamming weight of $\mathbf{p}$ is clearly the product of the Hamming weights of the component codewords, i.e.,

$$\mathcal{W}(\boldsymbol{p}) = \mathcal{W}(\boldsymbol{r})\mathcal{W}(\boldsymbol{c}). \tag{6.12}$$

Tolhuizen showed that any codewod with weight smaller than $w(d_r, d_c)$ is obvious (Theorem 1, [105]) (smaller or equal if $q = 2$ and both d_r and d_c are odd (Theorem 2, [105])). The term $\frac{1}{q-1}$ in (6.8) and (6.10) is due to the fact $(\lambda \boldsymbol{r}_i)(\boldsymbol{c}_j/\lambda)$ are equal for all nonzero $\lambda \in \mathbb{F}_q$.

A generalization of Tolhuizen's result to input-output weight enumerators is given in the following theorem.

Theorem 6.1. *Let $\mathcal{R}$ and $\mathcal{C}$ be (n_r, k_r, d_r) and (n_c, k_c, d_c) linear codes over $\mathbb{F}_q$, respectively. Given the product code $\mathcal{P} = \mathcal{R} \times \mathcal{C}$, the exact IOWE for codewords with output Hamming weight $1 < h < h_o$ is given by*

$$O_{\mathcal{P}}(w, h) = \frac{1}{q-1} \sum_{i|w} \sum_{j|h} O_{\mathcal{R}}(i, j) O_{\mathcal{C}}(w/i, h/j), \tag{6.13}$$

where the sum extends over all factors i and j of w and h respectively, and h_o is given by (6.9).

Proof. Let $\boldsymbol{p} \in \mathcal{P}$ be a rank-one codeword; then there exists a codeword $\boldsymbol{r} \in \mathcal{R}$ and a codeword $\boldsymbol{c} \in \mathcal{C}$ such that $\boldsymbol{p}_{i,j} = \boldsymbol{r}_i \boldsymbol{c}_j$. The $k_r k_c$ submatrix of information symbols in $\boldsymbol{p}$ could be constructed from the information symbols in $\boldsymbol{c}$ and $\boldsymbol{r}$ by (6.11) for $1 \leq i \leq k_r$ and $1 \leq j \leq k_c$. It thus follows that the input weight of $\boldsymbol{p}$ is the product of the input weights of $\boldsymbol{c}$ and $\boldsymbol{r}$ while its output (total) weight is given by (6.12). Since all codewords with weights $h < h_o$, are rank-one codewords, the theorem follows. □

These results show that both the weight enumerators and the input-output weight enumerators of product code low-weight codewords are determined by the constituent code low-weight enumerators. This is not the case for larger weights, where the enumerators of $\mathcal{P}$ are not completely determined by the enumerators of $\mathcal{R}$ and $\mathcal{C}$ [105].

It is important to note the number of rank-one low-weight codewords is very small, as shown by the following corollary regarding Reed-Solomon (RS) product codes.

Corollary 6.2. *Let $\mathcal{C}$ be an (n, k, d) Reed-Solomon code over $\mathbb{F}_q$. The weight enumerator of the product code $\mathcal{P} = \mathcal{C} \times \mathcal{C}$ has the following properties,*

$$
E_{\mathcal{P}}(h) = \begin{cases} 1, & h = 0 \\ (q-1)\left(\binom{n}{d}\right)^2, & h = d^2 \\ 0, & d^2 < h < d(d+1) \end{cases} \tag{6.14}
$$

Proof. Let us apply (6.8). From the maximum distance separable (MDS) property of RS codes, $d = n - k + 1$ and $n < q$. It follows that $w(d, d) = d(d + 1)$. Also $E_{\mathcal{C}}(d) = (q - 1)\binom{n}{d}$. The first obvious codeword of nonzero weight has weight d^2. The next possible nonzero obvious weight is $d(d + 1)$ which is $w(d, d)$. $\qquad\square$

Example 6.1. Let us consider the $\mathcal{C}(7, 5, 3)$ RS code. The number of codewords of minimum weight is $E_{\mathcal{C}}(d) = 245$. The complete IOWE function of $\mathcal{C}$ is equal to (see Corollary 3.4):

$$
\begin{aligned}
\mathbb{O}_{\mathcal{C}}(X, Y) = 1 \ &+ \ 35XY^3 + 140X^2Y^3 + 70X^3Y^3 + 350X^2Y^4 + 700X^3Y^4 \\
&+ \ 175X^4Y^4 + 2660X^3Y^5 + 2660X^4Y^5 + 9170X^4Y^6 + 266X^5Y^5 \\
&+ \ 3668X^5Y^6 + 12873X^5Y^7.
\end{aligned}
$$

Let $\mathcal{P}$ be the square product code $\mathcal{P} = \mathcal{C} \times \mathcal{C}$. The minimum distance of $\mathcal{P}$ is $d_p = 9$. By (6.8), its multiplicity is $E_{\mathcal{P}}(d_p) = 8575$. By applying Theorem 6.1, the input-output weight enumerator for codewords in $\mathcal{P}$ with output weight $d_p = 9$ is given by

$$
\text{Coeff}(\mathbb{O}_{\mathcal{P}}(X, Y), Y^9) = 175X + 1400X^2 + 700X^3 + 2800X^4 + 2800X^6 + 700X^9. \tag{6.15}
$$

By Corollary 6.2, there are no codewords in $\mathcal{P}$ with either weight 10 or 11. No information is available for larger codeword weights $12 \le w \le 49$. $\qquad\diamond$

The following theorem shows that rank-one codewords of a product code maintain the multiplicity property.

Theorem 6.3. *If the codes $\mathcal{C}$ and $\mathcal{R}$ have the multiplicity property and $\mathcal{P} = \mathcal{R} \times \mathcal{C}$ is their product code, then the subcode constituting of the rank-one codewords in $\mathcal{P}$ has this property.*

Proof. It follows from Theorem 6.1 that, for $h \leq h_o$

$$
\begin{aligned}
I_{\mathcal{P}}(h) &= \frac{1}{q-1} \sum_{w=1}^{k_r k_c} \frac{w}{k_r k_c} \sum_{i|w} \sum_{j|h} O_{\mathcal{R}}(i,j) O_{\mathcal{C}}(w/i, h/j) \\
&= \frac{1}{q-1} \sum_{j|h} \sum_{i=1}^{k_r} \frac{i}{k_r} O_{\mathcal{R}}(i,j) \sum_{t=1}^{k_c} \frac{t}{k_c} O_{\mathcal{C}}(t, h/j) \\
&= \frac{1}{q-1} \frac{h}{n_r n_c} \sum_{j|h} E_{\mathcal{R}}(j) E_{\mathcal{C}}(h/j) \\
&= \frac{h}{n_p} E_{\mathcal{P}}(h),
\end{aligned}
$$

which proves the assertion. □

6.3 Average IOWE of Product Codes

In the previous section, we have shown how to exactly compute the product code IOWE, for low weight codewords. For higher codeword weights, it is very hard to find the exact enumerators for an arbitrary product code over $\mathbb{F}_q$.

In this section, we will relax the problem of finding the exact enumerators, and we will focus on the computation of average weight enumerators over an ensemble of proper concatenated schemes. To do this:

1. We will represent a product code as a concatenated scheme with a row-by-column interleaver. Two representations will be introduced. The first one is the typical serial interpretation of a product code, while the second one is a less usual parallel construction.

2. We will replace the row-by-column interleavers of the schemes by uniform interleavers [9], acting as the average of all possible interleavers. To do this, we will introduce and discuss uniform interleavers for codes over $\mathbb{F}_q$.

3. We will compute the average enumerator for these concatenated schemes, which coincide with the scheme enumerators if random interleavers were used instead of row-by-column ones.

A code constructed using a random interleaver is no longer a rectangular product code. However, as we shall see, the average weight enumerator gives a very good approximation of the exact weight enumerator of the product code. This will confirm the experimental results by Hagenauer *et al.* that the error performance of linear product codes did not differ much if the row-column interleaver is replaced with a random interleaver [52, Sec. IV B]. We also confirm that numerically in Section 6.7.

6.3.1   Representing a Product Code as a Concatenated Code

Let us first study the representation of a product code as a concatenated scheme with a row-by-column interleaver.

Construction 1

Given the (n_r, k_r, d_r) code $\mathcal{R}$, the augmented code $\mathcal{R}^{k_c}$ is obtained by independently appending k_c codewords of $\mathcal{R}$. The code $\mathcal{R}^{k_c}$ has codeword length $k_c n_r$ and dimension $k_r k_c$. Moreover, its IOWE function is given by

$$\mathbb{O}_{\mathcal{R}}^{k_c}(X,Y) \triangleq \mathbb{O}_{\mathcal{R}^{k_c}}(X,Y) \;\; = \;\; (\mathbb{O}_{\mathcal{R}}(X,Y))^{k_c}. \tag{6.16}$$

See Figure 6.1. The encoding process may be viewed as if we are first generating a codeword of $\mathcal{R}^{k_c}$, with length $k_c n_r$ symbols. The symbols of this codeword are read into an $k_c \times n_r$ matrix by rows and read out column by column. In other words, the symbols of the augmented codeword are interleaved by a *row-by-column* interleaver. Each column is then encoded into a codeword in $\mathcal{C}$. The augmented columns form a codeword in $\mathcal{P}$ of length $n_r n_c$.

Remark. An $(n_r n_c, k_r k_c, d_r d_c)$ product code $\mathcal{P} = \mathcal{R} \times \mathcal{C}$ is the serial concatenation of an $(k_c n_r, k_c k_r)$ outer code $\mathcal{R}^{k_c}$ with an $(n_c n_r, k_c n_r)$ inner code $\mathcal{C}^{n_r}$ through a row-by-column interleaver π with length $N = k_c n_r$. (Equivalently $\mathcal{P} = \mathcal{R} \times \mathcal{C}$ is the serial concatenation of an $(k_r n_c, k_r k_c)$ outer code $\mathcal{C}^{k_r}$ with an $(n_c n_r, k_r n_c)$ inner code $\mathcal{R}^{n_c}$ through a row-by-column interleaver with length $k_r n_c$ respectively.)

Construction 2

As an alternative, let the coordinates of a systematic product code be partitioned into four sets as shown in Figure 6.2. We can introduce the following parallel repre-

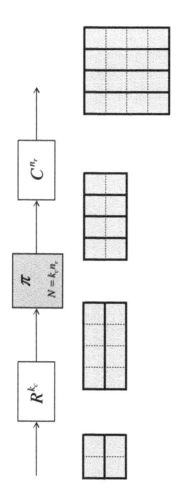

Figure 6.1: Construction 1: Serial concatenation.

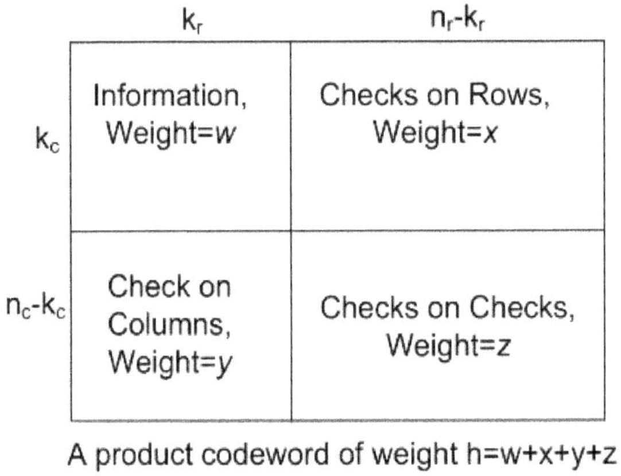

k_r	$n_r - k_r$
Information, Weight=w	Checks on Rows, Weight=x
Check on Columns, Weight=y	Checks on Checks, Weight=z

k_c (left label for first row), $n_c - k_c$ (left label for second row)

A product codeword of weight h=w+x+y+z

Figure 6.2: The four set partition of the coordinates of a product codeword used in Construction 2.

sentation.

Remark. An $(n_r n_c, k_r k_c, d_r d_c)$ product code can be constructed as follows (see Figure 6.3):

1. Parallel concatenate the $(n_r k_c, k_r k_c)$ code $\mathcal{R}^{k_c}$, with the $(n_c k_r, k_c k_r)$ code $\mathcal{C}^{k_r}$ through a row-by-column interleaver π_1 of length $N_1 = k_r k_c$.

2. Interleave the parity symbols generated by $\mathcal{R}^{k_c}$ with a row-by-column interleaver π_2 of length $N_2 = k_c(n_r - k_r)$.

3. Serially concatenate these interleaved parity symbols with the $(n_c(n_r - k_r), k_c(n_r - k_r))$ code $\mathcal{C}^{n_r - k_r}$.

6.3.2 Uniform Interleavers over $\mathbb{F}_q$

Given the two product code representations just introduced, we would like to substitute the row-by-column interleavers with uniform interleavers. In this section, we then investigate the uniform interleaver properties, when the interleaver is a symbol based interleaver and the symbols are in $\mathbb{F}_q$. The concept of uniform interleaver was introduced in [9] and [8] for binary vectors in order to study turbo codes: it is a probabilistic object acting as the average of all possible interleavers of the given length. In the binary case, the number of possible permutations of a vector of length L and Hamming weight w is $\binom{L}{w}$. Let us denote by $V(L, w)$ the probability that a specific vector is output by the interleaver when a vector of length L and input w is randomly interleaved. In this binary case we have

$$V(L, w) = \frac{1}{\binom{L}{w}}. \tag{6.17}$$

If $\boldsymbol{v}$ is a vector on $\mathbb{F}_q$ of length L and the frequency of occurrence of the q symbols is given by $l_0, l_1, ..., l_{q-1}$ respectively, then the number of possible permutations is given by the multinomial coefficient [109]

$$\frac{L!}{l_0! l_1! ... l_{q-1}!}.$$

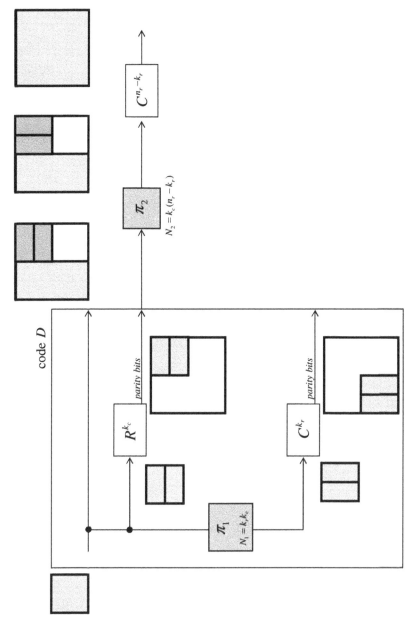

Figure 6.3: Construction 2: Parallel concatenation.

However, this requires the knowledge of the occurrence multiplicity of each of the q symbols in the permuted vector.

We introduce here the notion of *uniform codeword selector* (UCS). Let us suppose a specific vector of symbol weight w and length L is output from the interleaver corresponding to a certain interleaver input with the same weight. This vector is encoded by an (N, L) code $\mathcal{C}$ following the interleaver.

We assume that all the codewords of $\mathcal{C}$ with input weight w have equal probability of being chosen at the encoder's output. The UCS picks one of these codewords (with input weight w) at random. Thus the probability that a specific codeword is chosen by the UCS is

$$V(L, w) = \frac{1}{\sum_h O_{\mathcal{C}}(w, h)} = \frac{1}{(q-1)^w \binom{L}{w}}, \tag{6.18}$$

where $\sum_h O_{\mathcal{C}}(w, h)$ is the total number of codewords with input weight w. This is equivalent to a uniform interleaver over $\mathbb{F}_q$ which identifies codewords by their Hamming weights. It is noticed that for the binary case, the uniform interleaver (6.17) is equivalent to the UCS (6.18). The UCS has the property of preserving the cardinality of the resulting concatenated code.

6.3.3 Computing the Average Enumerators

Construction 1

Given the Construction 1 of Remark 6.3.1 and Figure 6.1, let us replace the row-by-column interleaver π of length $N = k_c n_r$ with a uniform interleaver over $\mathbb{F}_q$ of the same length. It is easy to show that the average IOWE function of the product code $\mathcal{P}$ is given by

$$\bar{\mathbb{O}}_{\mathcal{P}}(X, Y) = \sum_{w=0}^{k_c n_r} V(k_c n_r, w) \operatorname{Coeff}\left(\mathbb{O}_{\mathcal{R}}^{k_c}(X, Y), Y^w\right) \operatorname{Coeff}\left(\mathbb{O}_{\mathcal{C}}^{n_r}(X, Y), X^w\right). \tag{6.19}$$

The average weight enumerator function $\bar{\mathbb{E}}_{\mathcal{P}}(Y)$ can be computed from $\bar{\mathbb{O}}_{\mathcal{P}}(X, Y)$ by applying (6.6).

Construction 2

Given Construction 2 and Figure 6.3, let us replace the two row-by-column interleavers π_1 of length $N_1 = k_r k_c$ and π_2 of length $N_2 = k_c(n_r - k_r)$, with two uniform interleaver overs $\mathbb{F}_q$ of length N_1 and N_2, respectively.

We begin by finding the *partition weight enumerator* (PWE) of the code $\mathcal{D}$ resulting from the parallel concatenation of $\mathcal{R}^{k_c}$ with $\mathcal{C}^{k_r}$. We have:

$$\bar{\mathbb{P}}_{\mathcal{D}}(W, X, Y) = \sum_{w=0}^{k_c k_r} V(k_r k_c, w)$$
$$\text{Coeff}\left(\mathbb{R}_{\mathcal{R}}^{k_c}(W, X), W^w\right) \text{Coeff}\left(\mathbb{R}_{\mathcal{C}}^{k_r}(W, Y), W^w\right) W^w, \tag{6.20}$$

where $\bar{P}_{\mathcal{D}}(w, x, y)$ is the number of codewords in the parallel concatenated code with weights w, x and y in the partitions constituting of information symbols, checks on rows and checks on columns respectively, and is given by

$$\bar{\mathbb{P}}_{\mathcal{D}}(W, X, Y) = \sum_{w=0}^{k_c k_r} \sum_{x=0}^{k_c(n_r-k_r)} \sum_{y=0}^{k_r(n_c-k_c)} \bar{P}_{\mathcal{D}}(w, x, y) W^w X^x Y^y. \tag{6.21}$$

(Note that $\bar{\mathbb{R}}_{\mathcal{D}}(W, X) = \bar{\mathbb{P}}_{\mathcal{D}}(W, X, X)$ gives the average IRWE function of a punctured product code with the checks on checks deleted.)

The partition weight enumerator function of the product code $\mathcal{P}$ is then given by

$$\bar{\mathbb{P}}_{\mathcal{P}}(W, X, Y, Z) = \sum_{x=0}^{k_c(n_r-k_r)} V(k_c(n_r - k_r), x)$$
$$\text{Coeff}\left(\mathbb{R}_{\mathcal{C}}^{n_r-k_r}(X, Z), X^x\right) \text{Coeff}\left(\bar{\mathbb{P}}_{\mathcal{D}}(W, X, Y), X^x\right) X^x. \tag{6.22}$$

The PWE, $\bar{P}_{\mathcal{P}}(w, x, y, z)$, enumerates the codewords with a weight profile shown in Figure 6.2 and is given by expanding the PWE function $\bar{\mathbb{P}}_{\mathcal{P}}(W, X, Y, Z)$ as follows,

$$\bar{\mathbb{P}}_{\mathcal{P}}(W, X, Y, Z) = \sum_{w=0}^{k_c k_r} \sum_{x=0}^{k_c(n_r-k_r)} \sum_{y=0}^{k_r(n_c-k_c)} \sum_{z=0}^{(n_r-k_r)(n_c-k_c)} \bar{P}_{\mathcal{P}}(w, x, y, z) W^w X^x Y^y Z^z. \tag{6.23}$$

It follows that the average IRWE function of $\mathcal{P}$ is $\bar{\mathbb{R}}_{\mathcal{P}}(X, Y) = \bar{\mathbb{P}}_{\mathcal{P}}(X, Y, Y, Y)$. Consequently, the IOWE function $\bar{\mathbb{O}}_{\mathcal{P}}(X, Y)$ can be obtained via (6.5) and the weight enumerator function $\bar{\mathbb{E}}_{\mathcal{P}}(Y)$ via (6.6). By using (6.18), the cardinality of the code given by $\bar{E}_{\mathcal{P}}(Y)$ is preserved to be $q^{k_r k_c}$.

6.4 Merging Exact and Average Enumerators into Combined Enumerators

The results in the previous section are now combined with those of Section 6.2 reflecting our knowledge of the exact IOWE of product codes for low weights. Let h_o be defined as in (6.8). We introduce a complete IOWE which is equal to:

- the exact IOWE for $h < h_o$;
- the average IOWE for $h \geq h_o$:

$$\tilde{\mathbb{O}}_{\mathcal{P}}(X,Y) = \sum_{w=0}^{k_r k_c} \sum_{h=0}^{n_r n_c} \tilde{O}_{\mathcal{P}}(w,h) X^w Y^h, \tag{6.24}$$

such that

$$\tilde{O}_{\mathcal{P}}(w,h) = \begin{cases} O_{\mathcal{P}}(w,h), & h < h_o \\ \bar{O}_{\mathcal{P}}(w,h), & h \geq h_o \end{cases}, \tag{6.25}$$

where $O_{\mathcal{P}}(w,h)$ is given by Theorem 6.1, while $\bar{O}_{\mathcal{P}}(w,h) = \text{Coeff}(\tilde{\mathbb{O}}_{\mathcal{P}}(X,Y), X^w Y^h)$ is derived as in Section 6.3.3. We will call $\tilde{\mathbb{O}}_{\mathcal{P}}(X,Y)$ the *combined input-output weight enumerator* (CIOWE) of $\mathcal{P}$. The corresponding *combined weight enumerator function* $\tilde{\mathbb{E}}_{\mathcal{P}}(Y)$ can be computed by (6.6).

Let us now discuss some properties of the CIOWE. Let $W(\mathcal{C}) \triangleq \{h : E_{\mathcal{C}}(h) \neq 0\}$ be the set of weights h, such that there exists at least one codeword $\mathbf{c} \in \mathcal{C}$ with weight h. Observe that the weight of a product codeword $\mathbf{p} \in \mathcal{P}$ is simultaneously equal to the sum of the row weights and to the sum of the column weights. We define an integer h a *plausible* weight of $\mathbf{p} \in \mathcal{P}$, if h could be simultaneously partitioned into n_c integers restricted to $W(\mathcal{R})$ and into n_r integers restricted to $W(\mathcal{C})$.

Note however, that not all plausible weights are necessarily in $W(\mathcal{P})$.

Theorem 6.4. *Suppose $\mathcal{P} = \mathcal{R} \times \mathcal{R}$, (the row code $\mathcal{R}$ is the same as the column code $\mathcal{C}$), then the set of weights with a nonzero coefficient in the average weight enumerator of $\mathcal{P}$ derived by either (6.19) or (6.22) are plausible weights for the product code.*

Proof. The set of plausible weights of a product code is the set of weights h such the coefficients of Y^h in both $(\mathbb{E}_\mathcal{C}(Y))^{n_r}$ and $(\mathbb{E}_\mathcal{R}(Y))^{n_c}$ is nonzero. When $\mathcal{R} = \mathcal{C}$, it suffices to show that for any weight h if the coefficient of Y^h in $\bar{\mathbb{E}}_\mathcal{P}(Y)$ is nonzero, then it is also nonzero in $(\mathbb{E}_\mathcal{C}(Y))^{n_r}$.

For Construction 1, let $\bar{\mathbb{E}}_\mathcal{P}(Y)$ be the average weight enumerator derived from (6.19) by $\bar{\mathbb{E}}_\mathcal{P}(Y) = \bar{\mathbb{O}}_\mathcal{P}(1, Y)$. Since all output weights that appear in $\bar{\mathbb{O}}_\mathcal{P}(1, Y)$ are obtained from Coeff $(\mathbb{O}_\mathcal{C}^{n_r}(X, Y), X^w)$ then, by (6.16), they have nonzero coefficients in $(\mathbb{E}_\mathcal{C}(Y))^{n_r}$ and we are done.

For Construction 2, let $\bar{\mathbb{E}}_\mathcal{P}(Y)$ be the average weight enumerator derived from (6.22) by $\bar{\mathbb{E}}_\mathcal{P}(Y) = \bar{\mathbb{P}}_\mathcal{P}(Y, Y, Y, Y)$. Let $\Upsilon(W, Y) = \text{Coeff}\left(\bar{\mathbb{P}}_\mathcal{D}(W, X, Y), X^x\right)$. From (6.20), it follows that any exponent with a nonzero coefficients in $\Upsilon(Y, Y)$ also has a nonzero coefficient in $\mathbb{R}_\mathcal{C}^{k_r}(Y, Y)$ or equivalently $\mathbb{E}_\mathcal{C}^{k_r}(Y)$. Similarly if $\Upsilon'(X, Z) = \text{Coeff}\left(\mathbb{R}_\mathcal{C}^{n_r - k_r}(X, Z), X^x\right) X^x$, then any exponent with a nonzero coefficient in $\Upsilon'(Y, Y)$ also has a nonzero exponent in $\mathbb{E}_\mathcal{C}^{n_r - k_r}$. It follows from (6.22) that any exponent with a nonzero coefficient in $\bar{\mathbb{E}}_\mathcal{P}(Y)$ also has a nonzero coefficient in $\mathbb{E}_\mathcal{C}^{n_r - k_r}\mathbb{E}_\mathcal{C}^{k_r}$ and we are done. $\qquad\square$

In [106], the authors approximated the weight enumerator of the product code by a binomial distribution for *all* weights greater than h_o. Our approach has the advantage that only *plausible* weights appear in the combined enumerators of the product code.

6.5   Split Weight Enumerators of Linear Codes

As seen in the previous section, deriving the CIOWE of the product code requires the knowledge of the IRWE of the component codes. In this section we discuss the weight enumerators of some codes which are typically used for product code schemes. In particular, we show closed form formulas for the IRWE of Hamming, extended Hamming, Reed-Solomon codes. To do this, it is sometimes easier to work with the split weight enumerator (SWE, see definition in Section 6.1) of the dual code. The connection between the IRWE of a code and its dual was established in [114]. The following theorem gives a simplified McWilliams identity relating the SWE of a linear code with that of its dual code in terms of Krawtchouk polynomials.

Theorem 6.5. *Let C be an (n,k) linear code over $\mathbb{F}_q$ and $C^\perp$ be its dual code. Let $A(\alpha,\beta)$ and $A^\perp(\alpha,\beta)$ be the SWEs of C and $C^\perp$ respectively for an (n_1, n_2) partition of their coordinates, then*

$$A^\perp(\alpha,\beta) = \frac{1}{|C|} \sum_{w=0}^{n_1} \sum_{v=0}^{n_2} A(w,v)\mathcal{K}_\alpha(w,n_1)\mathcal{K}_\beta(v,n_2),$$

such that for $\beta = 0,1,\ldots,\gamma$, $\mathcal{K}_\beta(v,\gamma) = \sum_{j=0}^{\beta} \binom{\gamma-v}{\beta-j}\binom{v}{j}(-1)^j(q-1)^{\beta-j}$ is the Krawtchouk polynomial.

Proof. By a straightforward manipulation of the Macwilliams identity for the split weight enumerator [74, Chapter 5, (52)][62], it follows that for linear codes and $r = q-1$,

$$\mathbb{A}^\perp(X,Y) = \frac{1}{|C|}(1+rX)^{n_1}(1+rY)^{n_2}\mathbb{A}\left(\frac{1-X}{1+rX},\frac{1-Y}{1+rY}\right)$$

which is equivalent to

$$\mathbb{A}^\perp(X,Y) = \frac{1}{|C|}\sum_{w=0}^{n_1}\sum_{v=0}^{n_2} A(w,v)$$
$$(1-rX)^{n_1-w}(1-X)^w(1-rY)^{n_2-v}(1-Y)^v, \qquad (6.26)$$

where $\mathbb{A}(X,Y)$ and $\mathbb{A}^\perp(X,Y)$ are the SWE functions of C and $C^\perp$ respectively. Observing that for a positive integer γ and $0 \le \beta \le \gamma$, $(1-rY)^{\gamma-v}(1-Y)^v = \sum_{\beta=0}^{\gamma}\mathcal{K}_\beta(v,\gamma)Y^\beta$ is the generating function for the Krawtchouk polynomial [74, Chapter 5, (53)] and that $A^\perp(\alpha,\beta)$ is the coefficient of $X^\alpha Y^\beta$ in the right-hand side of (6.26) the result follows. $\qquad \square$

By observing that the roles of the input and the redundancy are interchanged in the code and its dual, we have:

Corollary 6.6. *The IRWEs of C and $C^\perp$ are related by*

$$R^\perp(\alpha,\beta) = \frac{1}{|C|}\sum_{v=0}^{n-k}\sum_{w=0}^{k} R(w,v)\mathcal{K}_\beta(w,k)\mathcal{K}_\alpha(v,n-k).$$

6.5.1 Hamming and Simplex Codes

The IRWE function of systematic Hamming codes could be derived by observing that they are the dual code of simplex codes [74, 73]. A recursive equation for evaluating the IRWE of Hamming codes was given in [96]. The following theorem gives a closed form formula for the IRWE function of Hamming codes in terms of Krawtchouk polynomials.

Theorem 6.7. *The IRWE of* $(2^m - 1, 2^m - m - 1, 3)$ *(systematic) Hamming codes is*

$$R_H(\alpha, \beta) =$$
$$\frac{1}{2^m} \left(\sum_{w=1}^{m} \binom{m}{w} \mathcal{K}_\beta(w, m) \mathcal{K}_\alpha(2^{m-1} - w, 2^m - m - 1) + \binom{m}{\beta} \binom{2^m - m - 1}{\alpha} \right).$$

Proof. By observing that the IRWEF of the $(2^m - 1, m, 2^{m-1})$ simplex code is

$$\mathbb{R}_s(X, Y) = 1 + \sum_{w=1}^{m} \binom{m}{w} X^w Y^{2^{m-1} - w}.$$

Using Corollary 6.6 and observing that $\mathcal{K}_\beta(0, m) = \binom{m}{\beta}$, we obtain the result.   □

6.5.2 Extended Hamming and Reed-Muller Codes

Extended Hamming codes were studied in [19], where it was shown they possess the multiplicity property, and closed-form formulas for their input-output multiplicity were provided. The following theorem shows a closed expression for their IRWE function in terms of Krawtchouk polynomials.

Theorem 6.8. *A closed form formula for the IRWE of the* $(2^m, 2^m - m - 1, 4)$ *extended Hamming codes is*

$$R_{EH}(\alpha, \beta) = \frac{1}{2^{m+1}} \left(\sum_{w=1}^{m} \binom{m+1}{w} \mathcal{K}_\beta(w, m+1) \, \mathcal{K}_\alpha(2^{m-1} - w, 2^m - m - 1) \right.$$
$$\left. + \binom{m+1}{\beta} \binom{2^m - m - 1}{\alpha} \left(1 + (-1)^{\alpha+\beta} \right) \right).$$

Proof. By observing that the extended Hamming codes are the duals of the $(2^m, m + 1, 2^{m-1})$ first-order Reed-Muller (RM) codes whose IRWE function could be shown to be

$$\mathbb{R}(X,Y) = 1 + X^{m+1}Y^{2^m-m-1} + \sum_{\alpha=1}^{m} \binom{m+1}{\alpha} X^{\alpha}Y^{2^m-1-\alpha}.$$

By Corollary 6.6 and observing that $\mathcal{K}_\beta(\gamma, \gamma) = \binom{\gamma}{\beta}(-1)^\beta$ the result follows.   □

Note that the WE of extended Hamming (EH) codes could also be derived from that of Hamming (H) codes by using the well-known relation [74], $E_{EH}(h) = E_H(h) + E_H(h-1)$ if h is even and is zero otherwise. It follows that

$$R_{EH}(\alpha, \beta) = \begin{cases} R_H(\alpha, \beta) + R_H(\alpha, \beta - 1), & \alpha + \beta \text{ is even} \\ 0, & \text{otherwise} \end{cases}. \tag{6.27}$$

6.5.3 Reed-Solomon Codes

Reed-Solomon codes are maximum distance separable (MDS) codes [74]. We have proved the following theorem (c.f., Theorem 3.3).

Theorem 6.9. *The SWE of MDS codes is given by*

$$A^T(w_1, w_2) = E(w_1 + w_2) \frac{\binom{n_1}{w_1}\binom{n_2}{w_2}}{\binom{n}{w_1+w_2}}.$$

It follows that the IRWE of an (n, k) systematic RS code is given by:

$$R_{RS}(\alpha, \beta) = E(\alpha + \beta) \frac{\binom{k}{\alpha}\binom{n-k}{\beta}}{\binom{n}{\alpha+\beta}}.$$

6.6   IRWE of Binary Images of Product Reed-Solomon Codes

Recently, new techniques for decoding Reed-Solomon codes beyond half-the-minimum distance were derived in [49], and algebraic soft-decision algorithms were proposed (c.f.,

Chapter 4 and Chapter 5). In this section we derive a number of results on RS product codes and their binary image.

Given the product of Reed-Solomon codes defined over a field of characteristic two, it is often the case that the binary image of the code is transmitted over a binary-input channel. The performance would thus depend on the binary weight enumerator of the component RS codes, which, as explained in Section 2.2, depends on the basis used to represent the 2^m-ary symbols as bits. The weight enumerator for the average binary image of codes, defined over finite fields of characteristic two, can be derived by assuming a binomial distribution of the bits in the nonzero symbols (c.f (2.9)). Let $\mathcal{C}_b$ denote the binary image of an (n, k) code $\mathcal{C}$ which is defined over the finite field $\mathbb{F}_{2^m}$. Let $\mathbb{E}_{\mathcal{C}}(Y)$ be the weight enumerator function of $\mathcal{C}$. Then the average weight enumerator of the (nm, km) code $\mathcal{C}_b$ is given by

$$\bar{\mathbb{E}}_{\mathcal{C}_b}(Y) = \mathbb{E}_{\mathcal{C}}(\Psi(Y)), \tag{6.28}$$

where $\Psi(Y) = \frac{1}{2^m-1}((1+Y)^m - 1)$ is the generating function of the bit distribution in a nonzero symbol. We assume that the distribution of the nonzero bits in a nonzero symbol follows a binomial distribution and that the nonzero symbols are independent. If the coordinates of the code $\mathcal{C}$ are split into p partitions, then there is a corresponding p-partition of the coordinates of $\mathcal{C}_b$, where each partition in $\mathcal{C}_b$ is the binary image of a partition in $\mathcal{C}$.

Let $\mathcal{P}_b$ denote the ensemble of binary images of the product code. By Theorem 3.14, the average partition weight enumerator of $\mathcal{P}_b$ could be derived as in the following lemma.

Lemma 6.10. *Let $\mathbb{P}_{\mathcal{P}}(W, X, Y, Z)$ be the PWE function of a code $\mathcal{P}$ defined over $\mathbb{F}_{2^m}$. The average PWE of the binary image $\mathcal{P}_b$ is*

$$\bar{\mathbb{P}}_{\mathcal{P}_b}(W, X, Y, Z) = \mathbb{P}_{\mathcal{P}}\left(\Psi(W), \Psi(X), \Psi(Y), \Psi(Z)\right).$$

Corollary 6.11. *If $\tilde{\mathbb{R}}_{\mathcal{P}}(X, Y)$ is the combined IRWE of the (n_p, k_p) product code $\mathcal{P}$*

defined over $\mathbb{F}_{2^m}$, *then the combined IRWE of its binary image is*

$$\tilde{\mathbb{R}}_{\mathcal{P}_b}(X,Y) = \tilde{\mathbb{R}}_{\mathcal{P}}(\Psi(X), \Psi(Y)),$$

where

$$\Psi(X) = \frac{1}{2^m - 1}((1 + X)^m - 1)$$

and

$$\tilde{\mathbb{R}}_{\mathcal{P}_b}(X,Y) = \sum_{x=0}^{k_p m} \sum_{y=0}^{n_p m - k_p m} R_{\mathcal{P}_b}(x,y)X^x Y^y.$$

Note this same formula does not hold in the case of the IOWE. However, the binary IOWE could be derived from the binary IRWE by using (6.5).

6.7   Numerical Results

In this section we show some numerical results supporting our theory. The combined input output enumerators of some product codes are investigated in Section 6.7.1. Analytical bounds to ML performance are computed and discussed in Section 6.7.2. Hamming codes, extended Hamming codes and Reed-Solomon codes are considered as constituent codes.

6.7.1   Combined Input-Output Weight Enumerators

Example 6.2. Let us consider the $(8,4)$ extended Hamming code. From Theorem 6.8, its IOWE function is

$$\mathbb{O}_{EH}(X,Y) = 1 + 4XY^4 + 6X^2Y^4 + 4X^3Y^4 + X^4Y^8.$$

Let us now study the $(8,4)^2$ square product code. By applying (6.19) we can derive the average weight enumerator function obtained with the serial concatenated repre-

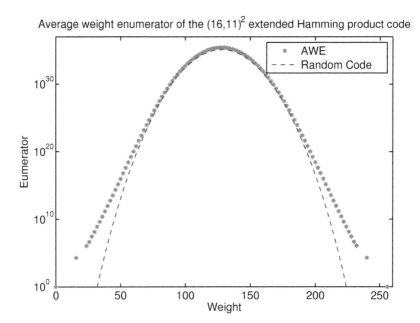

Figure 6.4:   The combined weight enumerator of the $(16, 11)^2$ extended Hamming product code is compared with that of a random binary code of the same dimension.

sentation. By rounding to the nearest integer, we obtain:

$$\mathbb{E}_P(Y) = 1 \;+\; 3Y^8 + 27Y^{12} + 107Y^{16} + 604Y^{20} + 3153Y^{24} + 13653Y^{28} + 30442Y^{32}$$
$$+ \;\; 13653Y^{36} + 3153Y^{40} + 604Y^{44} + 107Y^{48} + 27Y^{52} + 3Y^{56} + Y^{64}.$$

By (6.22), we can derive the average weight enumerator function obtained with the parallel concatenated representation. We obtain:

$$\mathbb{E}_P(Y) = 1 \;+\; 2Y^8 + 26Y^{12} + 98Y^{16} + 568Y^{20} + 3116Y^{24} + 13780Y^{28} + 30353Y^{32}$$
$$+ \;\; 13780Y^{36} + 3116Y^{40} + 568Y^{44} + 98Y^{48} + 26Y^{52} + 2Y^{56} + Y^{64}.$$

(For space limitations we do not show the IOWE functions.) Note that all codewords are of *plausible* weights as expected from Theorem 6.4. It could be checked that in both cases, the cardinality of the code (without rounding) is preserved to be 2^{16}. In general, the parallel representation gives more accurate results than the serial one, and will be used for the remaining results in this chapter.

For low-weight codewords, we can compute the exact IOWE. By Theorem 6.1, the exact IOWE of the product code for weights less than $h_o = 24$ is equal to

$$\mathbb{O}_P(X,Y) = 1 + 16XY^{16} + 48X^2Y^{16} + 32X^3Y^{16} + 36X^4Y^{16} + 48X^6Y^{16} + 16X^9Y^{16}.$$

It follows that the combined weight enumerator function for this product code is

$$\tilde{\mathbb{E}}_P(Y) = 1 \;+\; 196Y^{16} + 3116Y^{24} + 13781Y^{28} + 30353Y^{32} + 13781Y^{36}$$
$$+ \;\; 3116Y^{40} + 568Y^{44} + 98Y^{48} + 26Y^{52} + 2Y^{56} + Y^{64}.$$

A symmetric weight enumerator of the component codes implies a symmetric one for the product code. Thus, by the knowledge of the exact coefficients of exponents less than 24, $\tilde{\mathbb{E}}_P(Y)$ could be improved by setting the coefficients of Y^{54}, Y^{52} and Y^{56} to be zero and adjusting the coefficients of the middle exponents such that the cardinality of the code is preserved. We obtain:

$$\tilde{\mathbb{E}}'_{\mathcal{P}}(Y) = 1 \;\; + \;\; 196Y^{16} + 3164Y^{24} + 13995Y^{28} + 30824Y^{32}$$
$$+ \;\; 13995Y^{36} + 3164Y^{40} + 196Y^{48} + Y^{64}. \tag{6.29}$$

In this case, the exact weight enumerator can be found by exhaustively generating the 65536 codewords of the product code, and it is equal to:

$$\mathbb{E}_{\mathcal{P}}(Y) = 1 \;\; + \;\; 196Y^{16} + 4704Y^{24} + 10752Y^{28} + 34230Y^{32}$$
$$+ \;\; 10752Y^{36} + 4704Y^{40} + 196Y^{48} + Y^{64}.$$

It can be verified that the combined weight enumerator (6.29) gives a very good approximation of this exact weight enumerator.   ◇

Example 6.3. The combined weight enumerator of the extended Hamming product code $(16, 11)^2$, computed by applying (6.20) and (6.22), is depicted in Figure 6.4. It is observed that for medium weights, the distribution is close to that of random codes, which is given by

$$E(w) = q^{-(n_p - k_p)} \binom{n_p}{w} (q - 1)^w,$$

except that only plausible weights exist.   ◇

Example 6.4. The combined symbol weight enumerator of the $(7, 5)^2$ RS product codes over $\mathbb{F}_8$, computed by applying (6.20) and (6.22), is shown in Figure 6.5. It can be observed that the weight enumerator approaches that of a random code over $\mathbb{F}_8$ for large weights. The average binary weight enumerator of the $(147, 75)$ binary image, obtained by applying Corollary 6.11, is shown in Figure 6.6. It is superior to a random code at low weights and then, as expected, approaches that of a binary random code at larger weights.   ◇

6.7.2   Maximum-Likelihood Performance

In this section, we investigate product code performance. The combined weight enumerators are used to compute the Poltyrev bound [87], which gives tight analytical

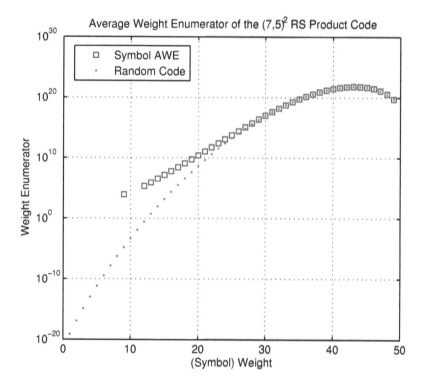

Figure 6.5:   The combined symbol weight enumerator of the $(7,5)^2$ Reed-Solomon product code is compared with that of a random code over $\mathbb{F}_8$ with the same dimension.

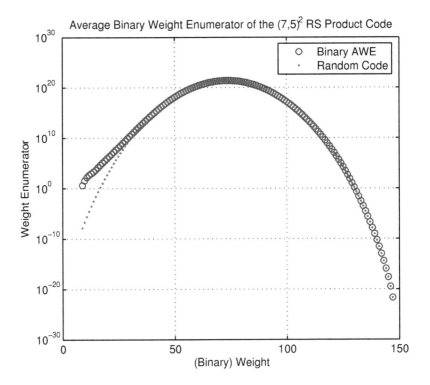

Figure 6.6: The combined binary weight enumerator of the binary image of the $(7,5)^2$ Reed-Solomon product codes is compared with that of a random binary code with the same dimension.

bounds to maximum-likelihood performance at both high and low error rates. For proper comparison, truncated union bound approximation and simulation results are also considered.

Example 6.5. The codeword error rate (CER) and the bit error rate (BER) performance of two Hamming product codes ($(7,4)^2$ and $(31,26)^2$) are shown in Figure 6.7. We have depicted:

- The Poltyrev bounds on ML performance (P on the plots), obtained by using the combined weight enumerator computed via (6.22).

- The truncated union bound (L on the plots), approximating the ML performance at low error rates, and computed from the minimum distance term via (6.7).

- The simulated performance of iterative decoding (S on the plots), corresponding to 15 iterations of the BCJR algorithm on the constituent codes trellises ([52],[19]).

By looking at the results, we can observe that:

- The combined weight enumerators derived in this chapter, in conjunction with the Poltyrev bound, provide very tight analytical bounds on the performance of maximum-likelihood decoding also at low SNRs (where the truncated union bound does not provide useful information).

- For the $(7,4)^2$ code the exact enumerator can be exhaustively computed, and the exact Poltyrev bound is shown in the figure. It is essentially identical to the bound computed with the combined weight enumerator.

- The ML analytical bounds provide very useful information also for iterative decoding performance. In fact, the penalty paid by iterative decoding with respect to ideal ML decoding is very limited, as shown in the figure (feedback coefficients for weighting the extrinsic information and improve iterative decoding has been employed, as explained in [19]).

◇

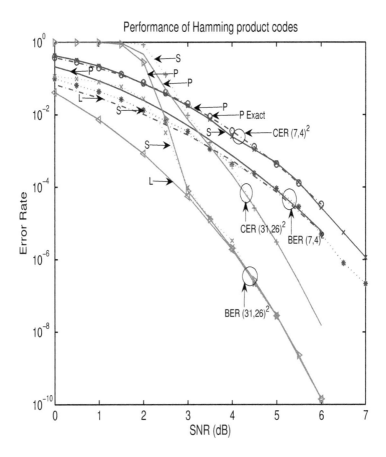

Figure 6.7:   CER and BER performance of some Hamming product codes for soft-decision decoding over AWGN channel.
The Poltyrev bound P, and the truncated union bound approximation L, are compared to simulated performance of iterative decoding S. For the $(7,4)^2$ code, the Poltyrev bound computed with the exact weight enumerator is also reported.

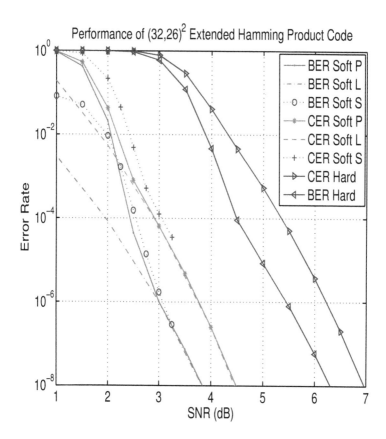

Figure 6.8: CER and BER performance of the $(32, 26)^2$ extended Hamming product code for soft-decision and hard-decision decoding over AWGN channel.
The Poltyrev bound P and the truncated union bound approximation L are compared to simulated performance of iterative decoding S.

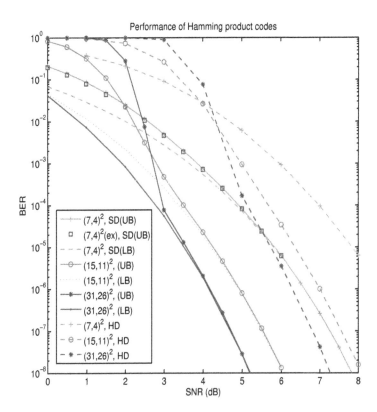

Figure 6.9: BER performance of Hamming product codes over AWGN channel. Bounds for both soft-decision (SD) and (HD) hard-decision decoding are shown. The Poltyrev upper bound (UB) and the truncated union bound approximation (LB) are used for SD, while the Poltyrev bound for the BSC is used for HD.

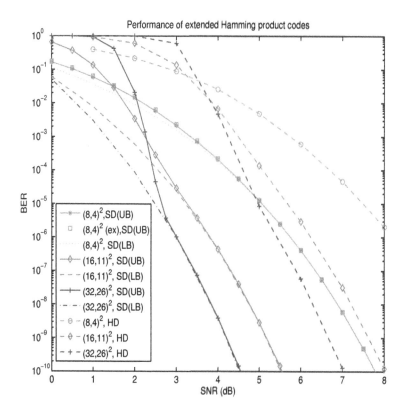

Figure 6.10: BER performance of extended Hamming product codes over AWGN channel.
Bounds for both soft-decision (SD) and (HD) hard-decision decoding are shown. The Poltyrev upper bound (UB) and the truncated union bound approximation (LB) are used for SD, while the Poltyrev bound for the BSC is used for HD.

Example 6.6. The performance of the extended Hamming product code $(32, 26)^2$ is investigated in Figure 6.8. Also in this case, the tightness of the bounds is demonstrated, for both the CER and the BER. With the aid of the Poltyrev bound for the BSC channel, hard ML bounds have also been plotted. It is shown that soft ML decoding on the AWGN channel offers more than 2 dB coding gain over hard ML decoding.
◇

Example 6.7. In Figure 6.9 and Figure 6.10, the performance of soft and hard ML decoding of various Hamming and extended Hamming codes are studied and compared. As expected, the EH product codes show better performance than Hamming product codes of the same length due to their larger minimum distance and lower rate. (For the $(7, 4)^2$ Hamming product code and the $(8, 4)^2$ extended Hamming product code, it is observed that the bounds using our combined weight enumerator overlapped with ones using the exact weight enumerators, which can be calculated exhaustively in these cases.)
◇

It is well known that the sphere packing bound provides a lower bound to the performance achievable by a code with given code-rate and codeword length [108]. The discrete-input further limitation occurring when using a given PSK modulation format was addressed in [14]. The distance of the code performance from this theoretical limit can be used an indicator of the code goodness.

Example 6.8. The performance of the binary image of some Reed-Solomon product codes, for both soft and hard decoding, are investigated in Figure 6.11, where the Poltyrev bound has been plotted. As expected, soft decoding has about 2 dB of gain over hard decoding. It can be observed that these product codes have good performance at very low error rates (BER lower than 10^{-9}), where no error floor appears. Let us consider, for example, the $(15, 11)^2$ RS product code, corresponding to a $(900, 484)$ binary code. By looking at the Poltyrev bound plotted in Figure 6.11, this code achieves a BER=10^{-10} for a signal-to-noise ratio $\gamma \simeq 2.2$ dB. By computing the PSK sphere packing bound for this binary code, we obtain a value of about 1.9 dB for BER=10^{-10}. This means that this RS product code is within 0.3 dB from the theoretical limit, which is a very good result at these low error rates.
◇

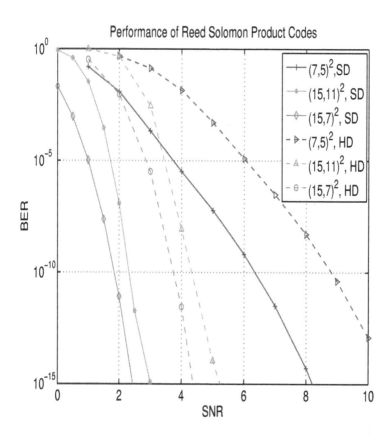

Figure 6.11: BER of some Reed-Solomon product codes over the AWGN channel. The performance bounds are plotted for both soft-decision (SD) and hard-decision (HD) decoding, using the combined weight enumerators.

6.8   Conclusion

The average weight enumerators of product codes were studied in this chapter. The problem was relaxed by considering proper concatenated representations, and assuming random interleavers over $\mathbb{F}_q$ instead of row-by-column interleavers. The exact IOWE for low-weight codewords were also derived by extending Tolhuizen results. By combining exact values and average values, a complete combined weight enumerator was computed. This enables us to study the ML performance of product codes at both low and high SNRs by applying the Poltyrev bound. The computation of average enumerators requires knowledge of the constituent code enumerators. Closed form formulas for the input redundancy enumerators of some popular codes were shown. The binary weight enumerator of ensemble of RS product codes was also derived.

The combined weight enumerators of Hamming and Reed-Solomon product codes were numerically computed and discussed. Using the combined enumerators, tight bounds on the ML performance of product codes over AWGN channels were derived by using the Poltyrev bounds. The tightness of the bounds were demonstrated by comparing them to both truncated union bound approximations and simulation results.

In particular, Reed-Solomon product codes show excellent performance. Reed-Solomon codes are widely used in wireless, data storage, and optical systems due to their burst-error-correction capabilities. The presented techniques allow to analytically estimate Reed-Solomon product codes performance, and show they are very promising as Shannon-approaching solutions down to very low error rates without error floors. This suggests the search for low-complexity soft decoding algorithms for Reed-Solomon codes as a very important research area in the near future.

Chapter 7

Algebraic List Decoding of Reed-Solomon Product Codes

> *Progress lies not in enhancing what is, but in advancing toward what will be.*
>
> —Khalil Gibran

In Chapter 6, we analyzed the performance of maximum-likelihood decoding of linear product codes. Product codes were introduced by Elias [38], who also proposed decoding them by iteratively decoding the component codes. They are widely used in data storage and satellite broadcast systems. Reed-Solomon (RS) product codes are product codes where the component codes are Reed-Solomon codes. A number of soft iterative decoding techniques have been devised for them [91, 5]. Maximum-likelihood performance analysis of Reed-Solomon product codes for both hard-decision and soft-decision decoding show the potential of devising improved polynomial time algorithms for decoding them [26].

We briefly refresh the definition of a product code. Assume that $\mathcal{R}$ and $\mathcal{C}$ are linear codes with parameters $(n_r, v_r + 1, d_r)$ and $(n_c, v_c + 1, d_c)$. The product code $\mathcal{P} = \mathcal{R} \times \mathcal{C}$ is defined as the set of all two-dimensional arrays such that each row of any array in $\mathcal{P}$ is a codeword of $\mathcal{R}$ and each column is a codeword of $\mathcal{C}$. It is well known that $\mathcal{P}$ is an $(n_p, v_p + 1, d_p) = (n_r n_c, (v_r + 1)(v_c + 1), d_r d_c)$ linear code. The rates of $\mathcal{R}$, $\mathcal{C}$ and $\mathcal{P}$ are $R_r = (v_r + 1)/n_r$, $R_c = (v_c + 1)/n_c$ and $R_p = R_r R_c$ respectively.

It is well known that the half-the-distance bound is not always attainable for product codes by iteratively decoding the component codes. For example, if the decoding algorithms for the row and column component codes are capable of correcting $(d_r - 1)/2$ and $(d_c - 1)/2$ errors respectively, and an error rectangular block of $((d_r - 1)/2 + 1) \times ((d_c - 1)/2 + 1)$ occurs, iterative decoding fails although the number of errors will be less than or equal to $(d_r d_c - 1)/2$ if $d_r d_c \geq d_r + d_c + 3$. If $d_r = d_c = d$, iterative decoding will fail for this pattern if $d \geq 3$. For example, the product of two $(7, 3, 5)$ RS codes has a minimum distance of 25 and a half-the-minimum distance decoder will be capable of correcting up to 12 errors. However, iterative decoding fails to correct the pattern of 9 errors described above.

Conventional bounded distance decoding algorithms for the component Reed-Solomon codes can correct up to half-the-minimum distance of the code. A list decoder will return a list of codewords with the goal of having the transmitted codeword on the list [39, 117]. List decoding of Reed-Solomon codes with the Guruswami-Sudan algorithm can correct errors beyond half-the-minimum distance of RS codes. The Guruswami-Sudan algorithm spurred a lot of progress in the area of list decoding of algebraic codes. Algorithms such as Sudan [102], Guruswami-Sudan [49], Parvaresh-Vardy [82, 81], and Guruswami-Rudra [48], show that we can basically decode above the half-the-minimum distance of the code for some specific codes. In this work, we investigate the generalization of Guruswami-Sudan algorithm for RS product code. We will that see this generalization results in algorithms that can decode more than half-the-minimum distance for certain rates of a RS product code.

This chapter is organized as follows. In Section 7.1, we introduce some notation and show that a Reed-Solomon product can be represented as an evaluation of a bivariate polynomial. In Section 7.2, we propose and analyze an algebraic list-decoding algorithm for two-dimensional Reed-Solomon product codes. The list-decoding algorithm is based on the interpolation and factorization ideas of the Guruswami-Sudan algorithm for decoding Reed-Solomon codes. In Section 7.3, we study M-dimensional Reed-Solomon product codes and generalize our algorithm and its analysis for an arbitrary dimension M. We then, in Section 7.4, investigate decoding product Reed-Solomon codes as subfield subcodes of Reed-Muller codes by invoking the Pellikan-Wu algorithm for decoding Reed-Muller codes. We conclude this chapter in Section 7.6.

7.1   Reed-Solomon Product Codes

We first briefly review the Reed-Solomon codes. Let $\mathbb{D}(X) = \sum_{i=0}^{v} D_i X^i$ be a data polynomial over $\mathbb{F}_q[X]$. [1] Then an $(n, v+1, d)$ Reed-Solomon code is generated by evaluating the data polynomial $\mathbb{D}(X)$ at n distinct elements of the field forming a set called the *support* set of the code $S = \{\alpha_0, \alpha_1, \ldots, \alpha_{n-1}\} \subset \mathbb{F}_q$. The generated codeword is $\mathbb{D}(S) = (\mathbb{D}(\alpha_0), \mathbb{D}(\alpha_1), \ldots, \mathbb{D}(\alpha_{n-1}))$. For a Reed-Solomon code, $d = n - v$.

Recall the definition of a product of two codes $\mathcal{P} = \mathcal{R} \times \mathcal{C}$ given in the introduction. We show how a product of two RS codes can be generated by polynomial evaluation of a bivariate polynomial.

Theorem 7.1. *Let $\mathcal{R}$ and $\mathcal{C}$ be $(n_r, v_r + 1, d_r)$ and $(n_c, v_c + 1, d_c)$ RS codes, respectively. Let $\mathcal{R}$ and $\mathcal{C}$ be defined as an evaluation codes over the support sets $S_r = \{\alpha_0, \alpha_1, \ldots, \alpha_{n_r-1}\} \subset \mathbb{F}_q$ and $S_c = \{\beta_0, \beta_1, \ldots, \beta_{n_c-1}\} \subset \mathbb{F}_q$ respectively. Define evaluation map:*

$$
\begin{aligned}
\mathrm{ev}^2 : \mathbb{F}_q[X, Y] &\longrightarrow \mathbb{F}_q^{n_r n_c} \\
\mathbb{D}(X, Y) &\longmapsto (\mathbb{D}(\alpha_i, \beta_j) : (\alpha_i, \beta_j) \in S_r \times S_c).
\end{aligned}
$$

Then the Reed-Solomon product code $\mathcal{P}$ is an evaluation code defined by

$$
\mathcal{P} = \mathcal{R} \times \mathcal{C} = \mathrm{ev}^2(L)
$$

where $L = \{\mathbb{D} \in \mathbb{F}_q[X, Y] \; : \; \deg_X \mathbb{D} \leq v_r \text{ and } \deg_Y \mathbb{D} \leq v_c\}$

Proof. Let $\mathbb{D}(X, Y) = \sum_{i=0}^{v_r} \sum_{j=0}^{v_c} D_{i,j} X^i Y^j$ be a data polynomial. The cardinality of the code generated by bivariate polynomial evaluation is equal to the number of distinct data polynomials $\mathbb{D}(X, Y)$, $q^{(v_r+1)(v_c+1)}$, which is equal to the cardinality of $\mathcal{R} \times \mathcal{C}$.

---

[1] We replace the ubiquitous $k - 1$ with v.

It is thus sufficient to show that the generated code $\mathcal{P}$ is a subcode of the product code $\mathcal{R} \times \mathcal{C}$. Consider a codeword $\boldsymbol{p} \in \mathcal{P}$ such that $\boldsymbol{p}_{i,j} = \mathbb{D}(\alpha_i, \beta_j)$. The rth row is equal to $\boldsymbol{p}_{r,*} = \{\mathbb{D}(\alpha_0, \beta_r), \mathbb{D}(\alpha_1, \beta_r), \ldots, \mathbb{D}(\alpha_{n_r-1}, \beta_r)\}$ where

$$\mathbb{D}(\alpha_c, \beta_r) = \sum_{i=0}^{v_r} \sum_{j=0}^{v_c} D_{i,j}(\alpha_c)^i(\beta_r)^j \tag{7.1}$$

$$= \sum_{i=0}^{v_r} \left(\sum_{j=0}^{v_c} D_{i,j}(\beta_r)^j \right) (\alpha_c)^i.$$

Define $\gamma_{r,s} = \sum_{j=0}^{v_c} D_{s,j}(\beta_r)^j$ and the univariate polynomial $\mathbb{D}'_r(X) = \sum_{i=0}^{v_r} \gamma_{r,i} X^i$. It is then easy to see that $\boldsymbol{p}_{r,*}$ can be generated by evaluating the modified data polynomial $\mathbb{D}'_r(X)$ at the support set S_r; $\boldsymbol{p}_{r,*} = \{\mathbb{D}'_r(\alpha_0), \mathbb{D}'_r(\alpha_1), \ldots, \mathbb{D}'_r(\alpha_{n_r-1})\}$. This proves that $\boldsymbol{p}_{r,*} \in \mathcal{R}$.

Similarly, any column c can be generated by evaluating the modified data polynomial $\mathbb{D}''_c(Y) = \sum_{j=0}^{v_c} \delta_{c,j} Y^j$ at the support set S_c; $\boldsymbol{p}_{*,c} = \{\mathbb{D}''_c(\beta_0), \mathbb{D}''_c(\beta_1), \ldots, \mathbb{D}''_c(\beta_{n_c-1})\}$, where $\delta_{c,j} = \sum_{i=0}^{v_r} D_{i,j}(\alpha_c)^i$. Thus each column $\boldsymbol{p}_{*,c}$ is a codeword in $\mathcal{C}$.

Since each row is a codeword in $\mathcal{R}$ and each column is a codeword in $\mathcal{C}$, then $\mathcal{P}$ is a subcode of $\mathcal{R} \times \mathcal{C}$ and we are done.   □

We will denote an RS product code, defined in Theorem 7.1 by $\mathcal{P}(S_r, S_c, v_r, v_c, q)$. It is easy to confirm that the minimum distance of $\mathcal{P}$ is indeed $d_r d_c$. From the above proof each row is generated by $\mathbb{D}'_r(X)$ of degree at most v_r. Since this univariate polynomial has at most v_r zeros, it will evaluate to at least $n_r - v_r$ nonzero values if it is nonzero. This means that at least $n_r - v_r$ columns are nonzero. Each of these columns is evaluated by the polynomial $\mathbb{D}''_c(Y)$. Thus each of these nonzero columns has at least $n_c - v_c$ nonzero positions. Thus if $\boldsymbol{p}$ is nonzero, the number of the nonzero elements in $\boldsymbol{p}$ is at least $(n_r - v_r)(n_c - v_c)$ which is $d_r d_c$. This proves the following corollary.

Corollary 7.2. *The number of distinct zeros of the bivariate polynomial* $\mathbb{D}(X,Y) = \sum_{i=0}^{v_r} \sum_{j=0}^{v_c} D_{i,j} X^i Y^j$ *is at most* $n_r v_c + n_c v_r - v_c v_r$ *if* $v_r < n_r$ *and* $v_c < n_c$.

For the sake of our analysis, we will need a bound on the total number of zeros, counting with multiplicities, of a bivariate polynomial. We will start by generalizing definitions 4.2 and 4.3 to M dimensions.

Definition 7.1. The $(r_1, r_2, \ldots, r_M)$th Hasse derivative of an M-variate polynomial $\mathbb{Q}(X_1, X_2, \ldots, X_M)$ at $(\alpha_1, \alpha_2, \ldots, \alpha_M)$, is given by

$$Q'_{r_1, r_2, \ldots, r_M}(\alpha_1, \alpha_2, \ldots, \alpha_M)$$
$$= \text{Coeff}(\mathbb{Q}(X_1 + \alpha_1, X_2 + \alpha_2, \ldots, X_M + \alpha_M), X_1^{r_1} X_2^{r_2} \ldots X_M^{r_M})$$
$$= \sum_{i_1, \ldots, i_M} \binom{i_1}{r_1} \cdots \binom{i_M}{r_M} Q_{i_1, \ldots, i_M} \alpha_1^{i_1 - r_1} \ldots \alpha_M^{i_M - r_M}.$$

Definition 7.2. The M-variate polynomial $\mathbb{Q}(X_1, X_2, \ldots, X_M)$ passes through the point $(\alpha_1, \alpha_2, \ldots, \alpha_M)$ with multiplicity m (has a zero of multiplicity m at $(\alpha_1, \alpha_2, \ldots, \alpha_M)$) iff $\mathbb{Q}(X_1 + \alpha_1, X_2 + \alpha_2, \ldots, X_M + \alpha_M)$ does not contain any polynomial of degree strictly less than m;

$$Q'_{r_1, \ldots, r_M}(\alpha_1, \ldots, \alpha_M) = 0 \text{ for all } r_1, r_2, \ldots, r_M \text{ such that } 0 \leq \sum_{i=1}^{M} r_i < m.$$

Definition 7.3. The $(w_1, w_2, \ldots, w_M)$-weighted degree of the M-variate polynomial $\mathbb{Q}(X_1, X_2, \ldots, X_M) = \sum_{i_1, i_2, \ldots, i_M} Q_{i_1, i_2, \ldots, i_M} X_1^{i_1} X_2^{i_2} \ldots X_M^{i_M}$ is

$$\deg_{w_1, w_2, \ldots, w_M} \mathbb{Q}(X_1, X_2, \ldots, X_M) \triangleq \max\{i_1 w_1 + i_2 w_2 + \cdots + i_M w_M \ : \ Q_{i_1, i_2, \ldots, i_M} \neq 0\}.$$

Theorem 7.3. *The number of zeros (counting with multiplicities) of the nonzero bivariate polynomial $\mathbb{D}(X, Y)$ evaluated over $S_r \times S_c$, where $|S_r| = n_r$ and $|S_c| = n_c$, is at most $\deg_{n_c, n_r} \mathbb{D}(X, Y)$.*

Proof. Let $v_c = \deg_Y \mathbb{D}(X, Y)$ and $v_r = \deg_X \mathbb{D}(X, Y)$. For any $\alpha \in \mathbb{F}_q$, $\mathbb{D}(\alpha, Y)$ is either the all zero polynomial or a polynomial in Y with maximum degree v_c. Define $\mathcal{G} \triangleq \{\gamma : (X - \gamma) | \mathbb{D}(X, Y)\}$. [2] Assuming that for each $\gamma_i \in \mathcal{G}$, m_i is the largest integer

---
[2] $(X - \gamma) | \mathbb{D}$ means that $(X - \gamma)$ is a factor of $\mathbb{D}$; $(X - \gamma)$ divides $\mathbb{D}$.

that $(X - \gamma_i)^{m_i}$ divides $\mathbb{D}(X, Y)$ then we can rewrite $\mathbb{D}(X, Y)$ as follows

$$\mathbb{D}(X, Y) = \left(\prod_{i=1}^{|\mathcal{G}|} (X - \gamma_i)^{m_i} \right) \mathbb{B}(X, Y)$$

where $\mathbb{B}(\alpha, Y)$ is a nonzero polynomial for any $\alpha \in S_r$ and $\deg_Y \mathbb{B}(X, Y) = v_c$.

For any $\gamma_i \in \mathcal{G}$, let assume that $\mathbb{B}(\gamma_i, Y)$ is zero at $\{\beta_1, \beta_2, \ldots, \beta_u\}$ with multiplicity $\{r_1, r_2, \ldots, r_u\}$, respectively. Then the number of zeros of $\mathbb{D}(\gamma_i, Y)$ counting with multiplicity over $S_r \times S_c$ is

$$\sum_{j=1}^{u} (m_i + r_j) + (n_c - u)m_i \leq um_i + v_c + (n_c - u)m_i.$$

The term $(n_c - u)m_i$ is the contribution of the points that $\mathbb{B}(\gamma_i, \beta)$ is not zero. Also notice that $\sum_j r_j \leq v_c$. By observing that $\sum_i m_i \leq v_r$, the total number of zeros for all $\gamma_i \in \mathcal{G}$ is

$$\sum_{i=1}^{|\mathcal{G}|} (v_c + n_c m_i) \leq |\mathcal{G}| v_c + n_c v_r.$$

For any $\alpha \notin \mathcal{G}$, $\mathbb{D}(\alpha, Y)$ is nonzero so it has at most v_c many zeros. Thus, the total number of the zeros is upper bounded by

$$(n_r - |\mathcal{G}|) v_c + |\mathcal{G}| v_c + n_c v_r = n_r v_c + n_c v_r,$$

which is $\deg_{n_c, n_r} \mathbb{D}(X, Y)$.   □

By comparing Corollary 7.2 and Theorem 7.3, we note that the number of distinct zeros of $\mathbb{D}(X, Y)$ is less than the total number of zeros by at most $v_r v_c$ zeros.

7.1.1   Half-the-Minimum Distance Bound

As we mentioned in the introduction, conventional methods for decoding product codes are not guaranteed to correct any pattern of errors with a cardinality of at most half-the-minimum distance of the code. However, it is important to compare the decoding radius of any decoding algorithm to half-the-minimum distance of the code. For a RS product code with minimum distance d_p,

$$
\begin{aligned}
\frac{d_p/2}{n_p} &\approx \frac{(1 - R_c)(1 - R_r)}{2} \\
&= 1 - \frac{1 + (R_c + R_r) - R_c R_r}{2} \\
&\leq 1 - \sqrt{R_c + R_r - R_c R_r},.
\end{aligned}
\tag{7.2}
$$

The last inequality follows from the arithmetic-geometric mean inequality. In case $R_c = R_r = \sqrt{R_p}$, then

$$
\begin{aligned}
\frac{d_p/2}{n_p} &\approx 1 - \sqrt{R_p} - \frac{1 - R_p}{2} \\
&= 1 - \frac{1 + 2\sqrt{R_p} - R_p}{2}.
\end{aligned}
\tag{7.3}
$$

Thus

$$
\begin{aligned}
\frac{d_p/2}{n_p} &\leq 1 - \sqrt{2\sqrt{R_p} - R_p} \\
&= 1 - \sqrt[4]{4R_p}\sqrt{1 - \frac{\sqrt{R_p}}{2}}.
\end{aligned}
\tag{7.4}
$$

This means that any decoder for product Reed-Solomon codes with an asymptotic relative decoding radius of $1 - \sqrt[4]{4R_p}\sqrt{1 - \frac{\sqrt{R_p}}{2}}$ will always decode beyond half-the-minimum distance of the code.

For comparison purposes, one can observe that the correction capability of Reed-Solomon codes is much larger, mainly due to their larger minimum distance. The half-the-minimum distance bound for RS codes is

$$
\begin{aligned}
\frac{d_{RS}/2}{n_{RS}} &\approx \frac{1 - R_{RS}}{2} \\
&\leq 1 - \sqrt{R_{RS}},
\end{aligned}
\tag{7.5}
$$
$$
\tag{7.6}
$$

where the latter upper bound is the Guruswami-Sudan radius for correcting Reed-Solomon codes. Thus when comparing the correction capability of different decoding algorithms for different codes one has to take into account the minimum distance of these codes. Since RS codes are maximum distance separable codes, they have the largest distance when compared to other codes with the same parameters. In Figure 7.1, we show the bounds on the decoding radius for RS product codes given by (7.3) and (7.4). We compare them with the bounds on the decoding radius for RS codes given by (7.5) and (7.6).

7.2   Algebraic Decoding Algorithm

In this section, we propose an algebraic algorithm for decoding RS product codes and analyze its performance. The Guruswami-Sudan (GS) algorithm is an algebraic decoding algorithm for RS codes which are defined as univariate evaluation polynomials. For $(n, v + 1, d)$ RS codes, the GS algorithm interpolates a bivariate polynomial through n interpolation points, defined by the support set of the code and the received word, in a two-dimensional space where n is the length of the RS code. Bivariate polynomial interpolation is followed by polynomial factorization where all linear factors of the bivariate polynomial with degree at most v are candidates for evaluation polynomials. We refer the reader to Section 4.2 for more details on the GS algorithm. The GS decoding algorithm can also be generalized for soft-decision decoding as explained in Section 4.4. The questions we will attempt to answer in this section are,

- Can one find a good interpolation-factorization algorithm for decoding (two-dimensional) product codes?

- What is the expected decoding radius of this decoding algorithm?

- What is the expected list size?

Theorem 7.1, hints at a generalization of the GS algorithm to trivariate polynomials. Assume that the Reed-Solomon product code $\mathcal{P} = \mathcal{R} \times \mathcal{C}$ is defined as in Theorem 7.1. The received word is $\boldsymbol{y} = [y_{i,j}]$, for $(i, j) \in \{1, 2, \ldots, n_r\} \times \{1, 2, \ldots, n_c\}$, given that the codeword $\boldsymbol{p} \in \mathcal{P}$ is transmitted. The Hamming distance between $\boldsymbol{y}$ and $\boldsymbol{p}$ will be denoted by $d(\boldsymbol{y}, \boldsymbol{p})$. Our algorithm can be formulated as follows:

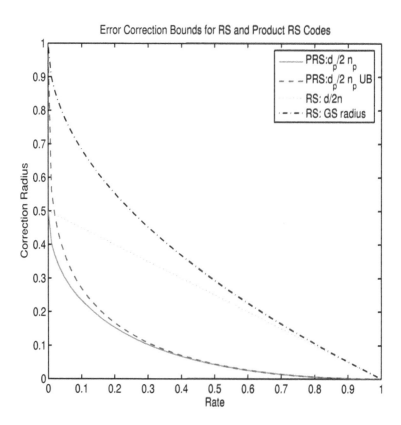

Figure 7.1: Error-correction capability for RS and RS product codes.
The half-the-distance bound for RS product codes $\frac{d_p/2}{n_p}$ (7.3) is compared with the
upper bound of (7.4). They are also compared to the classical decoding radius $\frac{d/2}{n}$ of
the component RS codes (7.5) and the Guruswami-Sudan decoding radius (7.6).

Algorithm 7.1. *Decoding of Product Reed-Solomon Codes. Let $\boldsymbol{y} \in \mathbb{F}_q^{n_p}$ be the received word when the codeword $\boldsymbol{p} \in \mathcal{P}(S_r, S_c, v_r, v_c, q)$ is transmitted.*

1. *Interpolate a trivariate polynomial $\mathbb{Q}(X, Y, Z)$ such that:*

   (a) *$\mathbb{Q} \neq 0$*

   (b) *$\mathbb{Q}(X, Y, Z)$ passes through the points $(\alpha_i, \beta_j, y_{i,j})$ with multiplicity m.*

   (c) *The $(n_c, n_r, n_c v_r + n_r v_c)$-weighted degree of $\mathbb{Q}(X, Y, Z)$ is less than Δ_m, where Δ_m is to be determined (Theorem 7.4).*

2. *Factorize $\mathbb{Q}(X, Y, Z)$ into irreducible factors. If $(Z - \mathbb{D}(X, Y)) | \mathbb{Q}(X, Y, Z)$, then $\hat{\boldsymbol{p}} = \mathrm{ev}^2(\mathbb{D}) = [\mathbb{D}(\alpha_i, \beta_j)]_{(\alpha_i, \beta_j) \in (S_r \times S_c)}$, is added to the list of candidates if*

   (a) *$\deg_X \mathbb{D}(X, Y) \leq v_r$ and $\deg_Y \mathbb{D}(X, Y) \leq v_c$*

   (b) *$\mathrm{d}(\hat{\boldsymbol{p}}, \boldsymbol{y}) \leq \tau_m$ where τ_m is the error-correction capability (determined by Theorem 7.7).*

This algorithm can be run in polynomial time in the length of the code n_p and the interpolation multiplicity m. As we will see, interpolating a polynomial amounts to solving a number of linear equations in a number of unknowns which are the coefficients of the polynomial. Thus it can be solved using Gaussian elimination or by a generalization of Koetter's interpolation algorithm or the Feng-Tzeng algorithm [71, 76]. The worst case complexity of the interpolation step is thus cubic in the number of unknowns (given by (7.9)). Finding the linear factors of this interpolated trivariate (or M-variate) polynomial can be done by a straightforward generalization of the Roth-Ruckenstein algorithm [95, 76] or other efficient factorization algorithms [119, 120]. The complexity of the algorithm is dominated by that of the factorization step.

The performance of the above algorithm depends on the choice of the interpolation multiplicity m. The larger the interpolation multiplicity m, the larger the decoding radius τ_m and the higher the computational complexity of the decoding algorithm. As m goes to infinity, the algorithm can correct any pattern of errors with a cardinality equal to its asymptotic decoding radius

$$\frac{\tau}{n_p} = 1 - \sqrt[6]{4 R_p}, \tag{7.7}$$

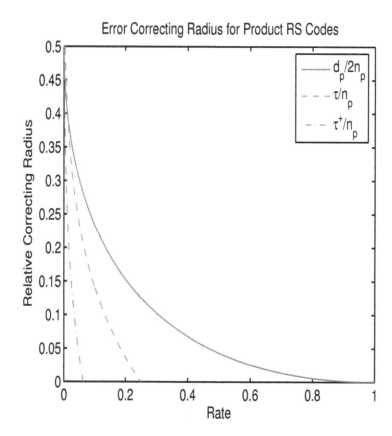

Figure 7.2: Decoding radii of different decoding algorithms for RS product codes. The half-the-distance bound for RS product codes $\frac{d_p/2}{n_p}$ (7.3) is compared with the decoding radius τ/n_p of Algorithm 7.1 given by (7.7) and the decoding radius τ^+/n_p of Algorithm 7.2 given by (7.8).

where it is assumed that $R_c = R_r = \sqrt{R_p}$, the length of the product code is n_p and its rate is R_p.

We stress that Algorithm 7.1 is not the only possible interpolation-factorization for decoding RS product codes. For example, suppose step 1c of Algorithm 7.1 is replaced by step 1c in the following

Algorithm 7.2. *Alternative Algorithm for Decoding Product Reed-Solomon Codes.*

1. *(a)* $\mathbb{Q} \neq 0$

    (b) $\mathbb{Q}(X,Y,Z)$ *passes through the points* $(\alpha_i, \beta_j, y_{i,j})$ *with multiplicity* m.

    (c) *The* $(1,0,v_r)$*-weighted degree of* $\mathbb{Q}(X,Y,Z)$ *is less than* Δ'_m *and the* $(0,1,v_c)$*-weighted degree of* $\mathbb{Q}(X,Y,Z)$ *is less than* Δ''_m *where* Δ'_m *and* Δ''_m *are to be determined.*

Then the error-correcting radius of the algorithm becomes

$$\frac{\tau^+}{n_p} = 1 - \sqrt[6]{16R_p}, \tag{7.8}$$

which is inferior to our proposed Algorithm 7.1, as seen from Figure 7.2. In the remaining of this section, we will analyze Algorithm 7.1 and prove that its asymptotic decoding radius is indeed given by (7.7).

7.2.1   Analysis of Algorithm 7.1

In step 1 of Algorithm 7.1, a trivariate polynomial $\mathbb{Q}(X,Y,Z) \in \mathbb{F}_q[X,Y,Z]$ is interpolated to pass through all the $(\alpha_i, \beta_j, y_{i,j})$ with multiplicity m.

Theorem 7.4. *There exist a nonzero trivariate polynomial* $\mathbb{Q}(X,Y,Z) \in \mathbb{F}_q[X,Y,Z]$ *such that* $\mathbb{Q}(X,Y,Z)$ *passes through all the* $(\alpha_i, \beta_j, y_{i,j})$, *for* $(i,j) \in \{1,2,\ldots,n_r\} \times \{1,2,\ldots,n_c\}$, *with multiplicity* m *and* $\deg_{n_c,n_r,n_c v_r + n_r v_c} \mathbb{Q}(X,Y,Z) \leq \Delta_m$ *where*

$$\Delta_m = \left\lceil m(n_r n_c) \sqrt[3]{(R_r + R_c)\left(1 + \frac{1}{m}\right)\left(1 + \frac{2}{m}\right)} \right\rceil.$$

Proof. The polynomial can be interpolated as long as the number of linear constraints imposed by step 1b of Algorithm 7.1 is strictly less than the number of unknowns. The unknowns are the coefficients of the monomials of $\mathbb{Q}(X, Y, Z)$ such that their weighted degree satisfy condition 1c of Algorithm 7.1. Let $N(\Delta)$ be the number of trivariate monomials whose $(n_c, n_r, n_c v_r + n_r v_c)$-weighted degree is at most Δ. $N(\Delta)$ can be lower bounded by the volume of a pyramid in $\mathbb{R}^3$ [20]. Considering the pyramid in Figure 7.3,

$$N(\Delta) > \frac{1}{6} \frac{\Delta^3}{n_r n_c (n_c v_r + n_r v_c)}.$$

From Definition 7.2, the number of constraints imposed by each point $(\alpha_i, \beta_j, y_{i,j})$ is equal to the number of solutions in nonnegative integers (r_1, r_2, r_3) to $0 \leq r_1 + r_2 + r_3 < m$ which is $\frac{m(m+1)(m+2)}{6}$. It follows that there exists a nonzero polynomial of weighted degree at most Δ that passes through all the points $(\alpha_i, \beta_j, y_{i,j})$ with multiplicity m if

$$N(\Delta) > n_r n_c \frac{m(m+1)(m+2)}{6}. \tag{7.9}$$

This implies the following condition

$$\deg_{n_c, n_r, n_c v_r + n_r v_c} \mathbb{Q}(X, Y, Z) \leq$$
$$\left\lceil m(n_r n_c) \sqrt[3]{\left(\frac{v_r}{n_r} + \frac{v_c}{n_c}\right)\left(1 + \frac{1}{m}\right)\left(1 + \frac{2}{m}\right)} \right\rceil, \tag{7.10}$$

and the theorem follows. $\qquad\square$

We know turn our attention to the factorization step of the algorithm. We will find a sufficient condition for a data polynomial $\mathbb{D}(X, Y)$ to be on the list output by the algorithm.

Theorem 7.5. *Let* $\boldsymbol{p} = (\mathbb{D}(\alpha_i, \beta_j) : (\alpha_i, \beta_j) \in S_r \times S_c)$ *and* $\boldsymbol{y}$ *the received word. Define* $\mathbb{H}(X, Y) \triangleq \mathbb{Q}(X, Y, \mathbb{D}(X, Y))$. *If*

$$\deg_{n_c, n_r} \mathbb{H}(X, Y) < m(n_r n_c - \mathrm{d}(\boldsymbol{y}, \boldsymbol{p})),$$

then $(Z - \mathbb{D}(X, Y))$ *is a factor of* $\mathbb{Q}(X, Y, Z)$.

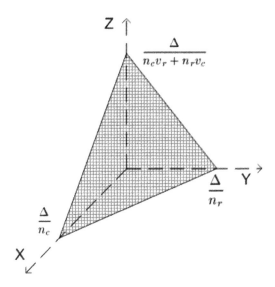

Figure 7.3: The number of monomials of maximum weighted degree Δ is lower bounded by the volume of this pyramid in $\mathbb{R}^3$.

Proof. From condition 1b of Algorithm 7.1 and Theorem 7.1, $\mathbb{H}(\alpha_i, \beta_j)$ is zero with multiplicity m for any (i,j) such that $y_{i,j} = p_{i,j}$. It follows that $\mathbb{H}(X,Y)$ has at least $m(n_r n_c - d(\boldsymbol{y}, \boldsymbol{p}))$ many zeros on $S_r \times S_c$. From Theorem 7.3, if the number of zeros of $\mathbb{H}(X,Y)$ becomes larger than $\deg_{n_c,n_r} \mathbb{H}(X,Y)$, then $\mathbb{H}(X,Y)$ is the zero polynomial.   □

Lemma 7.6. *The (n_c, n_r)-weighted degree of $\mathbb{H}(X,Y)$ is less than or equal to the $(n_c, n_r, n_c v_r + n_r v_c)$-weighted degree of $\mathbb{Q}(X,Y,Z)$.*

Proof. Assume that $X^i Y^j Z^\ell$ is a monomial of $\mathbb{Q}(X,Y,Z)$. When Z is substituted by $\mathbb{D}(X,Y)$, for this monomial we have

$$
\begin{aligned}
\deg_{n_c,n_r} X^i Y^j (\mathbb{D}(X,Y))^\ell &\leq \deg_{n_c,n_r} X^i Y^j (X^{v_r} Y^{v_c})^\ell \\
&\leq n_c i + n_r j + (n_c v_r + n_r v_c)\ell \\
&= \deg_{n_c,n_r,n_c v_r + n_r v_c} X^i Y^j Z^\ell.
\end{aligned}
$$

Therefore, the lemma is true for a general polynomial.   □

The following theorem gives a bound on the decoding radius of our list-decoding algorithm.

Theorem 7.7. *Assume we transmit a codeword $\boldsymbol{p} \in P(S_r, S_c, v_r, v_c, q)$ with row and column component code rates R_r and R_c respectively. Let $\boldsymbol{y} = [y_{i,j}]$ be the received word. If m is the interpolation multiplicity, then $\boldsymbol{p}$ can be efficiently list decoded from $\boldsymbol{y}$ if the Hamming distance between $\boldsymbol{y}$ and $\boldsymbol{p}$, $\tau_m = d(\boldsymbol{y}, \boldsymbol{c})$, is bounded by*

$$
\tau_m \leq \left\lfloor n_c n_r \left(1 - \sqrt[3]{(R_c + R_r)\left(1 + \frac{1}{m}\right)\left(1 + \frac{2}{m}\right)} \right) - \frac{1}{m} \right\rfloor.
$$

Proof. On one hand, by Theorem 7.5 and Lemma 7.6, $(Z - \mathbb{D}(X,Y))$ is a factor of the interpolated polynomial $\mathbb{Q}(X,Y,Z)$ if

$$
d(\boldsymbol{y}, \boldsymbol{p}) < n_r n_c - \frac{\deg_{n_c,n_r,n_c v_r + n_r v_c} \mathbb{Q}}{m}.
$$

On the other hand, to ensure that $\mathbb{Q}(X, Y, Z)$ exists and is nonzero, then by Theorem 7.4

$$\deg_{n_c, n_r, n_c v_r + n_r v_c} \mathbb{Q}(X, Y, Z) \leq \left\lceil m(n_r n_c) \sqrt[3]{\left(\frac{v_r}{n_r} + \frac{v_c}{n_c}\right)\left(1 + \frac{1}{m}\right)\left(1 + \frac{2}{m}\right)} \right\rceil.$$

By combining these two results the theorem follows. $\square$

Corollary 7.8. *For an interpolation multiplicity m, the error-correction radius τ_m is upper bounded by*

$$\tau_m \leq \left\lfloor n_p \left(1 - \sqrt[6]{4R_p} \sqrt[3]{\left(1 + \frac{1}{m}\right)\left(1 + \frac{2}{m}\right)}\right) - \frac{1}{m} \right\rfloor, \tag{7.11}$$

where R_p and n_p are the rate and length of the product code, respectively. The upper bound on the decoding radius is maximized when R_r is equal to R_c.

Proof. From the arithmetic-geometric mean inequality, $R_r + R_c \geq 2\sqrt{R_r R_c}$ with equality if $R_r = R_c = \sqrt{R_p}$. The result then follows directly from Theorem 7.7. $\square$

It thus follows that as the multiplicity m tends to infinity, the relative *asymptotic decoding radius* of the proposed algorithm is

$$\frac{\tau}{n_p} = \lim_{m \to \infty} \frac{\tau_m}{n_p} < 1 - \sqrt[3]{R_c + R_r}$$
$$\leq 1 - \sqrt[6]{4R_p}. \tag{7.12}$$

Remark. The list-decoding algorithm can correct *any pattern* of errors of cardinality greater than that of half-the-minimum distance decoder when $R_c + R_r \leq 0.22$ for sufficiently large m (Figure 7.4). In terms of R_p, it will be better than half-the-minimum distance if $R_p \leq 0.0121$. As we mentioned in the introduction, we do not know of a decoder that can correct *any pattern* of errors with a cardinality equal to that of half-the-minimum distance for RS product codes.

Although the product code has a rectangular (X, Y) support, the bound on the number of zeros of Theorem 7.3 depends on the total degree of the polynomial rather

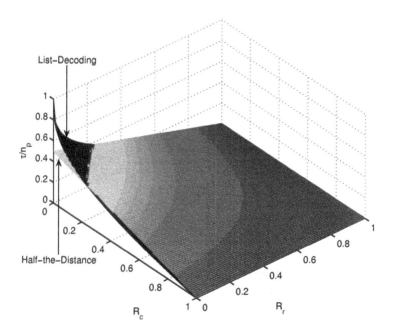

Figure 7.4: The $1 - \sqrt[3]{R_c + R_r}$ decoding radius and the half-the-distance bound.

than on the separate X and Y degrees. It follows that interpolating the polynomial $\mathbb{Q}(X, Y, Z)$ to have a triangular (X, Y) support as in Algorithm 7.1, rather than a rectangular (X, Y) support as in Algorithm 2.2, gives more monomials to work with. This is the main reason that the decoding radius of (7.7) is larger than that of (7.8).

The following theorem shows that the number of candidates on the decoding list of our proposed algorithm does not increase with the code length, n_p, or the alphabet size, q but rather on the rate of the code. For an interpolation multiplicity m we show that the list size L_m behaves like

$$L_m \propto R_p^{-1/3}. \tag{7.13}$$

Theorem 7.9. *For interpolating with a fixed multiplicity m, and for any received word $\mathbf{y} \in \mathbb{F}_q^{n_p}$, the candidate list size is upper bounded by*

$$L_m < \left\lceil m \sqrt[3]{\frac{1}{4R_p}\left(1+\frac{1}{m}\right)\left(1+\frac{2}{m}\right)} \right\rceil + 1. \tag{7.14}$$

Proof. The total number of candidate words on the list, counting plausible and implausible words, is upper bounded by the number of factors of $\mathbb{Q}(X,Y,Z)$ which are of the form $Z - \mathbb{D}(X,Y)$. This is upper bounded by the Z-degree of the polynomial $\mathbb{Q}(X,Y,Z)$. From Figure 7.3 and (7.10), we can see this can be upper bounded by

$$
\begin{aligned}
L_m \quad &< \quad \frac{\Delta}{n_c v_r + n_r v_c} \\
&\leq \quad m \sqrt[3]{\left(\frac{n_r n_c}{n_c v_r + n_r v_c}\right)^2 \left(1+\frac{1}{m}\right)\left(1+\frac{2}{m}\right)} \\
&\approx \quad m \sqrt[3]{\left(\frac{1}{R_c + R_r}\right)^2 \left(1+\frac{1}{m}\right)\left(1+\frac{2}{m}\right)} \\
&\leq \quad m \sqrt[3]{\frac{1}{4R_p}\left(1+\frac{1}{m}\right)\left(1+\frac{2}{m}\right)},
\end{aligned}
$$

where the last inequality follows from $\frac{1}{2}(R_c + R_r) \geq \sqrt{R_p}$ with equality if R_c is equal to R_p. $\qquad\square$

It is worth noting that the list size of the Guruswami-Sudan algorithm for decoding Reed-Solomon codes is bounded by [76].

$$L_m^{GS} \approx \left(m + \frac{1}{2}\right)\sqrt{\frac{1}{R}}. \tag{7.15}$$

The decoding algorithm with a smaller list size and a larger decoding radius is preferred.

7.3 Decoding M-dimensional Reed-Solomon Product Codes

A Reed-Solomon product code in M dimensions is an evaluation code defined by

$$
\begin{aligned}
\mathcal{P} &= \mathcal{C}_1 \times \mathcal{C}_2 \times \cdots \times \mathcal{C}_M \\
&= \mathrm{ev}^M(L),
\end{aligned}
$$

where $L = \{\mathbb{D} \in \mathbb{F}_q[X_1, X_2, \ldots, X_M] \;:\; \deg_{X_i} \mathbb{D} \le v_i \text{ for } i \in \{1, 2, \ldots, M\}\}$,

$$
\mathrm{ev}^M \;:\; \mathbb{F}_q[X_1, X_2, \ldots, X_M] \to \mathbb{F}_q^{\prod_{i=1}^M n_i}, \tag{7.16}
$$

and letting $S_1, S_2, \ldots, S_M$ to be the support sets along the M dimensions respectively,

$$
\begin{aligned}
&\mathbb{D}(X_1, X_2, \ldots, X_M) \;\mapsto\; \\
&\quad (\mathbb{D}(\alpha_1, \alpha_2, \ldots, \alpha_M) \;:\; (\alpha_1, \alpha_2, \ldots, \alpha_M) \in (S_1 \times S_2 \times \cdots \times S_M)).
\end{aligned}
$$

By a generalization of Theorem 7.1, one can show that a word along the ith dimension is a codeword in $\mathcal{C}_i$. If n_i, R_i and d_i denote the length, rate and minimum distance of the RS code $\mathcal{C}_i$, then for the product code $\mathcal{P}$, $n_p = \prod_{i=1}^M n_i$, $R_p = \prod_{i=1}^M R_i$ and $d_p = \prod_{i=1}^M d_i$. The half-the-distance bound will be given by

$$
\frac{d_p/2}{n_p} = \frac{\prod_{i=1}^M (n_i - v_i)}{2\, n_p} \approx \frac{\prod_{i=1}^M (1 - R_i)}{2} \tag{7.17}
$$

which is equal to

$$
\frac{d_p/2}{n_p} = \frac{\left(1 - \sqrt[M]{R_p}\right)^M}{2} \tag{7.18}
$$

if $R_1 = \cdots = R_M = \sqrt[M]{R_p}$.

7.3.1   The Decoding Algorithm

We start by giving a bound on the number of zeros, counting with multiplicities, of the multivariate polynomial $\mathbb{D}(X_1, X_2, \ldots, X_M)$, denoted by $\text{Zeros}\,[\mathbb{D}(X_1, X_2, \ldots, X_M)]$.

Theorem 7.10. *The number of zeros (counting with multiplicities) of the nonzero M-variate polynomial $\mathbb{D}(X_1, X_2, \ldots, X_M)$ evaluated over $S_1 \times S_2 \times \cdots \times S_M$, where $|S_i| = n_i$, $i \in \{1, 2, \ldots, M\}$, and $\deg_{X_i} \mathbb{D} = v_i$ is at most the $\left(\prod_{\substack{j \neq 1 \\ j \in \{1, \ldots, M\}}} n_j, \; \ldots, \prod_{\substack{j \neq M \\ j \in \{1, \ldots, M\}}} n_j \right)$-weighted degree of $\mathbb{D}(X_1, X_2, \ldots, X_M)$ which is*

$$\sum_{i=1}^{M} v_i \prod_{\substack{j \neq i \\ j \in \{1, \ldots, M\}}} n_j.$$

Proof. The proof follows by induction on M. By Theorem 7.3, it holds for $M = 2$. Now suppose it holds for M, then the number of zeros with multiplicities is at most

$$\text{Zeros}\,[\mathbb{D}(X_1, X_2, \ldots, X_M)] = \sum_{i=1}^{M} v_i \prod_{\substack{j \neq i \\ j \in \{1, \ldots, M\}}} n_j. \tag{7.19}$$

Now consider $\mathbb{D}(X_1, \ldots, X_M, X_{M+1})$. Let $\mathcal{G} = \{\gamma_i \in S_{M+1} \; : \; (X_{M+1} - \gamma_i)^{m_i} | \mathbb{D}\}$. We note that $\sum_{\gamma_i \in \mathcal{G}} m_i \leq v_{M+1}$. Let $\mathcal{G}' = S_{M+1} \setminus \mathcal{G}$. The number of zeros contributed by all $\gamma \in \mathcal{G}'$ is

$$(n_{M+1} - |\mathcal{G}|)\text{Zeros}[\mathbb{D}(X_1, \ldots, X_M)]. \tag{7.20}$$

Let

$$\mathbb{D}(X_1, \ldots, X_{M+1}) = \left(\prod_{i=1}^{|\mathcal{G}|} (X_{M+1} - \gamma_i)^{m_i} \right) \mathbb{B}(X_1, \ldots, X_{M+1}),$$

and for any $\gamma_i \in \mathcal{G}$, let $\mathbb{B}(X_1, \ldots, X_M, \gamma_i)$ be zero on u tubles of $(X_{1,j}, \ldots, X_{M,j})$ each with multiplicity r_j, then the number of such zeros is $\sum_{\gamma_i \in \mathcal{G}} \sum_{j=1}^{u} (m_i + r_j)$. The number of remaining zeros when $\mathbb{B}(X_1, \ldots, X_M, \gamma_i)$ is not zero is $\sum_{\gamma_i \in \mathcal{G}} (\prod_{k=1}^{M} n_k - u) m_i$. It follows the total number of zeros due to $\mathcal{G}$ is upper bounded by

$$|\mathcal{G}|\,\text{Zeros}[\mathbb{D}(X_1,\ldots,X_M)] + v_{M+1}\prod_{j=1}^{M} n_j. \tag{7.21}$$

By (7.19) and adding (7.20) to (7.21), one gets that

$$\text{Zeros}[\mathbb{D}(X_1,\ldots,X_{M+1})] \le \left(\sum_{i=1}^{M} v_i \prod_{\substack{j\neq i \\ j\in\{1,\ldots,M+1\}}} n_j\right) + \left(v_{M+1}\prod_{\substack{j\neq M+1 \\ j\in\{1,\ldots,M+1\}}} n_j\right).$$

Thus, $\text{Zeros}[\mathbb{D}(X_1,\ldots,X_{M+1})]$ is equal to the $\left(\prod_{\substack{j\neq 1 \\ j\in\{1,\ldots,M+1\}}} n_j,\ldots,\prod_{\substack{j\neq M+1 \\ j\in\{1,\ldots,M+1\}}} n_j\right)$-weighted degree of $\mathbb{D}(X_1, X_2,\ldots, X_{M+1})$. $\qquad\square$

We now generalize our decoding algorithm for M-dimensional Reed-Solomon product codes. For simplicity we will assume that an $(n, v + 1, d)$ RS code $\mathcal{C}$ with support set $S_c = \{\alpha_1,\ldots,\alpha_{n_c}\}$ is used as the component code along all M dimensions.

Algorithm 7.3. *Decoding of M-dimensional Product Reed-Solomon Codes. Let $y \in \mathbb{F}_q^{n_p}$ be the received word when the codeword $p \in \mathcal{P}$ is transmitted.*

1. *Interpolate an $(M + 1)$-variate polynomial $\mathbb{Q}(X_1, X_2,\ldots, X_M, Z)$ such that:*

   (a) $\mathbb{Q} \neq 0$

   (b) $\mathbb{Q}(X_1, X_2,\ldots, X_M, Z)$ *passes through all the points $(\alpha_{i_1}, \alpha_{i_2},\ldots, \alpha_{i_M},$ $y_{i_1,i_2,\ldots,i_M})$ with multiplicity m.*

   (c) *The $(n^{M-1},\ldots, n^{M-1}, Mn^{M-1}v)$-weighted degree of $\mathbb{Q}(X_1, X_2,\ldots, X_M, Z)$ is less than Δ_m, where Δ_m is to be determined (Theorem 7.12).*

2. *Factorize $\mathbb{Q}(X_1, X_2,\ldots, X_M, Z)$ into irreducible factors.*
   If $(Z - \mathbb{D}(X_1, X_2,\ldots, X_M))$ is a factor of $\mathbb{Q}(X_1, X_2,\ldots, X_M, Z)$, then $\hat{p} = \text{ev}^M \mathbb{D}(X_1, X_2,\ldots, X_M$ is added to the list of candidates if

   (a) $\deg_{X_i} \mathbb{D}(X_1, X_2,\ldots, X_M) \le v$ *for all $i = 1,\ldots, M$.*

(b) $\mathrm{d}(\hat{\boldsymbol{p}}, \boldsymbol{y}) \leq \tau_m$ *where τ_m is the error-correction capability (determined by Theorem 7.13).*

Similar to the two-dimensional case, Algorithm 7.3 can be run in polynomial time. Its complexity is dominated by that of the interpolation step. The complexity of the interpolation step is at most cubic in the number of the coefficients of the interpolated polynomial. The number of the coefficients can be shown to be bounded by $n^M \binom{m+M}{M+1}$ (see the proof of Theorem 7.12).

7.3.2   Analysis of the Algorithm

Define

$$\mathbb{H}(X_1, \ldots, X_M) \overset{\triangle}{=} \mathbb{Q}(X_1, \ldots, X_M, \mathbb{D}(X_1, \ldots, X_M)).$$

By Theorem 7.10, we can now give a bound on the number of zeros of $\mathbb{H}(X_1, \ldots, X_M)$.

Theorem 7.11. *Let $\mathbb{D}(X_1, \ldots, X_M)$ be defined as in Theorem 7.10, then the number of zeros of $\mathbb{H}(X_1, \ldots, X_M)$, counting with multiplicities is at most the*

$$\left(\prod_{\substack{j \neq 1 \\ j \in \{1, \ldots, M\}}} n_j, \; \ldots, \; \prod_{\substack{j \neq M \\ j \in \{1, \ldots, M\}}} n_j, \; \sum_{i=1}^{M} v_i \prod_{\substack{j \neq i \\ j \in \{1, \ldots, M\}}} n_j \right)$$

-weighted degree of $\mathbb{Q}$. If $n_i = n$ and $v_i = v$ for $i \in \{1, \ldots, M\}$, then the number of zeros of $\mathbb{H}$ is at most the $(n^{M-1}, \ldots, n^{M-1}, Mn^{M-1}v)$-weighted degree of $\mathbb{Q}$.

Proof. By Theorem 7.10, the number of zeros of $\mathbb{H}(X_1, \ldots, X_M)$ is upper bound by its $\left(\prod_{\substack{j \neq 1 \\ j \in \{1, \ldots, M\}}} n_j, \ldots, \prod_{\substack{j \neq M \\ j \in \{1, \ldots, M\}}} n_j \right)$-weighted degree, which in turn can be upper bounded by an upper bound on the weighted degree of the monomial $(X_1^{i_1} \ldots X_M^{i_M} (X_1^{v_1} \ldots X_M^{v_M})^{\ell})$ where ℓ is $\deg_Z \mathbb{Q}$ and the proof follows.   □

Theorem 7.12. *There exist a nonzero $(M+1)$-variate polynomial $\mathbb{Q}(X_1, X_2, \ldots, X_M, Z)$ $\in \mathbb{F}_q[X_1, X_2, \ldots, X_M, Z]$ such that $\mathbb{Q}(X_1, X_2, \ldots, X_M, Z)$ passes through all the points $(\alpha_{i_1}, \alpha_{i_2}, \ldots, \alpha_{i_M}, y_{i_1, i_2, \ldots, i_M})$, for $(i_1, i_2, \ldots, i_M) \in \{1, 2, \ldots, n\}^M$, with multiplicity m*

and the $\left(n^{M-1}, \ldots, n^{M-1}, Mn^{M-1}v\right)$-weighted degree of $\mathcal{Q} \leq \Delta_m$ where

$$\Delta_m = \left\lceil m \, n_p \sqrt[M+1]{M\frac{v}{n} \left(1 + \frac{1}{m}\right)\left(1 + \frac{2}{m}\right) + \cdots + \left(1 + \frac{M}{m}\right)} \right\rceil.$$

Proof. Let $N(\Delta)$ be the number of $(M + 1)$-variate monomials whose $\left(n^{M-1}, \ldots, n^{M-1}, Mn^{M-1}v\right)$-weighted degree is at most Δ. $N(\Delta)$ can be lower bounded by the volume of a pyramid in $\mathbb{R}^{M+1}$ [20] defined by the half planes

$$\{X_i \geq 0\}_{i=1}^{M} \,, \; Z \geq 0 \text{ and } \sum_{i=1}^{M} n^{M-1}X_i + Mn^{M-1}vZ \leq \Delta.$$

It follows that

$$\begin{aligned}
N(\Delta) \;>\; & \frac{1}{(M+1)!}\left(\frac{\Delta}{n^{M-1}}\right)^M \frac{\Delta}{Mn^{M-1}v} \\
=\; & \frac{1}{(M+1)!}\frac{\Delta^{M+1}}{Mn^{M^2}\frac{v}{n}}.
\end{aligned}$$

The number of linear constraints imposed by each interpolation point is the number of solutions in nonnegative integers a_i to $\sum_{i=1}^{M+1} a_i < m$ or equivalently $\sum_{i=1}^{M+2} a_i = m - 1$ which is $\binom{m+M}{M+1}$. As in the trivariate case, a solution to the interpolation problem exists if

$$n_p \binom{m + M}{M + 1} < N(\Delta).$$

This implies that

$$\Delta^{M+1} < n^{M(M+1)}m^{M+1}M\frac{v}{n}\left(1 + \frac{1}{m}\right)\left(1 + \frac{2}{m}\right) + \cdots + \left(1 + \frac{M}{m}\right),$$

and the result follows by noticing that $n_p = n^M$.   □

Theorem 7.13. *For an M-dimensional Reed-Solomon product code, Algorithm 7.3, with an interpolation multiplicity m, can correct any pattern of errors of cardinality at* *most*

$$\tau_m \leq$$

$$\left\lfloor n_p \left(1 - \sqrt[M(M+1)]{M^M R_p} \sqrt[M+1]{\left(1+\frac{1}{m}\right)\left(1+\frac{2}{m}\right)+\cdots+\left(1+\frac{M}{m}\right)}\right) - \frac{1}{m}\right\rfloor,$$

where R_p and n_p are the rate and length of the product code, respectively.

Proof. The proof is along the same lines of two-dimensional product codes. Let $\boldsymbol{p} = \mathrm{ev}^M \mathbb{D}$ and $\boldsymbol{y}$ be the received word. Then $\mathbb{H}(X_1,\ldots,X_M) \triangleq \mathbb{Q}(X_1,\ldots,X_M,\mathbb{D})$ has at least $m(n_p - \mathrm{d}(\boldsymbol{p},\boldsymbol{y}))$ zeros. By Theorem 7.10, it follows that $\mathbb{H}$ is the all zero polynomial and $(Z - \mathbb{D})|\mathbb{Q}$ if this number is greater than its $(n^{M-1},\ldots,n^{M-1})$-weighted degree. By Theorem 7.11, it follows that $(Z - \mathbb{D})$ is a factor of the interpolated polynomial $\mathbb{Q}$ if

$$\mathrm{d}(\boldsymbol{y},\boldsymbol{p}) < n_p - \frac{\deg_{n^{M-1},\ldots,n^{M-1},Mn^{M-1}v}\mathbb{Q}}{m}.$$

By Theorem 7.12 and letting $R \approx \frac{v}{n}$,

$$\mathrm{d}(\boldsymbol{y},\boldsymbol{p}) \leq \left\lfloor n_p \left(1 - \sqrt[M+1]{MR\left(1+\frac{1}{m}\right)\left(1+\frac{2}{m}\right)+\cdots+\left(1+\frac{M}{m}\right)}\right) - \frac{1}{m}\right\rfloor,$$

and the result follows. □

Corollary 7.14. *If the $R_1,\ldots,R_M$ are the rates of the component RS codes for an M-dimensional RS product code of rate $R_p = \prod_{i=1}^{M} R_i$, then, in the limit as the multiplicity m tends to infinity, the asymptotic relative decoding radius of the algorithm is*

$$\frac{\tau}{n_p} = \lim_{m\to\infty} \frac{\tau_m}{n_p}$$
$$\leq 1 - \sqrt[M+1]{R_1 + R_2 + \ldots + R_M}$$
$$\leq 1 - \sqrt[M(M+1)]{M^M R_p},$$

and the decoding radius is maximized when $R_1 = R_2 = \cdots = R_M$.

We finalize this section by generalizing the bound on the list size to the M-dimensional case.

Theorem 7.15. *The list returned by Algorithm 7.3 will have at most*

$$\left\lceil m^{M+1}\sqrt{\frac{1}{M^M R_p}\left(1+\frac{1}{m}\right)\left(1+\frac{2}{m}\right)\cdots\left(1+\frac{M}{m}\right)}\right\rceil + 1$$

codewords.

Proof. By the same arguments in Theorem 7.9, the size of the list can be upper bounded by $\deg_Z Q$. Thus, by Theorem 7.12,

$$L_m \ < \ \frac{\Delta_m}{Mn^{M-1}v}$$

$$< \ m\left(M\frac{v}{n}\right)^{\frac{-M}{M+1}}{}^{M+1}\sqrt{\left(1+\frac{1}{m}\right)\left(1+\frac{2}{m}\right)\cdots\left(1+\frac{M}{m}\right)},$$

which reduces to the desired result with $R_p = R^M$ and $R \approx \frac{v}{n}$.   □

For large m, the bound on the list size behaves like

$$L_m \ \propto \ mM^{\frac{-M}{M+1}} R_p^{\frac{-1}{M+1}}.$$

That means that it is decreasing with the number of dimensions, M, for a fixed product code rate R_p. For large M, the list size is almost proportional to $\frac{1}{M}$. The list size is also decreasing with the rate R_p for a fixed dimension M.

7.4   Decoding a Reed-Solomon Product Code as a Subcode of a Reed-Muller Code

A Reed-Muller code with M variables, of order r, denoted by $\mathrm{RM}_q(r, M)$ is an evaluation code defined by

$$\mathrm{RM}_q(r, M) = \mathrm{ev}^M(L')$$

where

$$L' = \{\mathbb{D} \in \mathbb{F}_q[X_1, X_2, \ldots, X_M] \ : \ \deg \mathbb{D} \le r\}$$

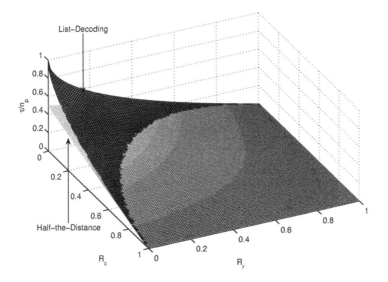

Figure 7.5: The $1 - \sqrt{R_c + R_r}$ decoding radius and half-the-distance bound

and $\deg \mathbb{D}$ is the total degree of $\mathbb{D}$. The evaluation map is similar to that of M-dimensional RS product codes (7.16)

$$\mathrm{ev}^M \; : \; \mathbb{F}_q[X_1, X_2, \ldots, X_M] \rightarrow \mathbb{F}_q^{q^M} \tag{7.22}$$

$$\mathbb{D}(X_1, X_2, \ldots, X_M) \longmapsto$$
$$(\mathbb{D}(\alpha_1, \alpha_2, \ldots, \alpha_M) \; : \; (\alpha_1, \alpha_2, \ldots, \alpha_M) \in (\mathbb{F}_q \times \mathbb{F}_q \times \cdots \times \mathbb{F}_q)).$$

If an M-dimensional RS product code is evaluated on $(\mathbb{F}_q \times \mathbb{F}_q \times \cdots \times \mathbb{F}_q)$ then its length is $n_p = q^M$. If the space of evaluated polynomials is

$$L = \{\mathbb{D} \in \mathbb{F}_q[X_1, X_2, \ldots, X_M] \; : \; \deg_{X_i} \mathbb{D} \leq v_i \text{ for } i \in \{1, 2, \ldots, M\}\},$$

then we will denote this code by $\mathrm{PRS}_q(v_1, \ldots, v_M)$. Then it is easy to see that it is a subcode of a Reed-Muller code

$$\mathrm{PRS}_q(v_1, \ldots, v_M) \subseteq \mathrm{RM}_q(v_1 + v_2 + \cdots + v_M, M). \tag{7.23}$$

Therefore, any algorithm used for decoding the RM code can be used for decoding the RS product code. From [86, 68] we know that the $RM_q(v_c + v_r, 2)$ is a subfield subcode of a generalized Reed-Solomon code over $\mathbb{F}_{q^2}$. With this observation, Pellikaan and Wu present a polynomial list-decoding algorithm for q-ary RM codes by invoking the list-decoding algorithm for Reed-Solomon codes. Thus, by decoding the generalized Reed-Solomon code using the Guruswami-Sudan algorithm [49] basically we can decode the RS product code.

Theorem 7.16 (Pellikaan and Wu [86]). *The Reed-Muller code $RM_q(r, M)$ can be efficiently list decoded with an error-correcting radius*

$$\tau < n \left(1 - \sqrt{1 - \frac{d}{n}} \right), \tag{7.24}$$

where d is the minimum distance of the q-ary Reed-Muller code of length n. When the rate is small, $r < q$, the minimum distance of $RM_q(r, M)$ is $d = (q - r)q^{M-1}$ and the decoding radius is

$$\tau < n \left(1 - \sqrt{\frac{r}{q}} \right). \tag{7.25}$$

Theorem 7.17. $\mathrm{PRS}_q(v_1, \ldots, v_M)$, *an M-dimensional RS product code evaluated over $\mathbb{F}_q^M$, can be list-decoded in polynomial time using the Pellikaan-Wu interpretation with a relative error-correcting radius of*

$$\frac{\tau}{n_p} < 1 - \sqrt{R_1 + R_2 + \cdots + R_M}$$

provided that $\sum_{i=1}^{M} R_i < 1$. In terms of R_p, the relative decoding radius is

$$\frac{\tau}{n_p} < 1 - \sqrt[2M]{M^M R_p},$$

provided that $R_p < M^{-M}$.

Proof. The proof follows by (7.23) and Theorem 7.16. The condition $r < q$ implies that $\sum_{i=1}^{M} R_i < 1$. Since

$$1 > \sum_{i=1}^{M} R_i \geq M \sqrt[M]{R_p},$$

this implies the condition $R_p < M^{-M}$ and that

$$\frac{\tau}{n_p} < 1 - \sqrt{M \sqrt[M]{R_p}}$$

and we are done. □

Corollary 7.18. *For a two-dimensional product code, the relative decoding radius with the Pellikaan and Wu interpretation is*

$$\begin{aligned} \frac{\tau}{n_p} \; &< \; 1 - \sqrt{R_1 + R_2} \\ &\leq \; 1 - \sqrt[4]{4R_p}. \end{aligned}$$

In Figure 7.5, we show the decoding region of Corollary 7.18 in terms of the rates of the component codes. It is worth comparing the result of Theorem 7.17 with that of Corollary 7.14. Both results are only valid for $R_p \leq M^{-M}$. This hints that the effective operating region of the algorithms go down exponentially in M. Also the gap between the decoding radii of Theorem 7.17 and Corollary 7.14 in the effective decoding region decreases as the M increases. In Figure 7.6, we compare the error-correcting capability using the Pellikaan-Wu interpretation with that of Algorithm 7.1 and the half-the-minimum distance bound. We observe that decoding product codes as subcodes of RM codes results in a larger error-correcting radius.

7.5 Discussion

We recall the argument that we had in Section 7.1.1, that we hope for an algorithm that can correct any pattern of errors beyond half-the-minimum distance. We showed

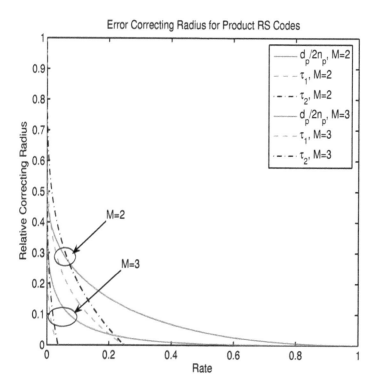

Figure 7.6: Error-correcting radii of list-decoding algorithms for two-dimensional and three-dimensional RS product codes.
The half-the-distance bound is denoted by $\frac{d_p}{2n_p}$. The decoding radii τ_1 is given by Corollary 7.14 and τ_2 is given by Theorem 7.17.

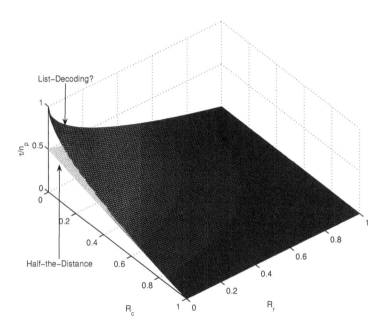

Figure 7.7:   The optimistic $1 - \sqrt{R_c + R_r - R_r R_c}$ decoding radius and the half-the-distance bound.

that a decoding algorithm with a relative decoding radius of $1 - \sqrt{R_c + R_r - R_r R_c}$ can successfully do so. In Figure 7.7, we also show that by comparing this bound to that of $d_p/2n_p$. Such an algorithm will exist if it is true that the RS product code $\mathcal{P}(S_r, S_c, v_r, v_c, q)$ is a subfield subcode of a generalized RS code over $\mathbb{F}_{q^2}$ with the same minimum distance of the product code, $(n_r - v_r)(n_c - v_c)$, length $n_r n_c$ and consequently dimension of $n_r v_c + n_c v_r - v_r v_c + 1$. By decoding the generalized RS code with the Guruswami-Sudan algorithm, the desired decoding radius can be achieved. To our knowledge it remains open whether this conjecture is true.

7.6   Conclusion

Product Reed-Solomon codes are widely used in data storage, optical and satellite communication systems. M-dimensional Reed-Solomon product codes can be regarded as an evaluation of an M-variate polynomial with constraints on its degrees. In this work, we proposed polynomial time algorithms for efficient list decoding of Reed-Solomon product codes.

The first algorithm is based on a generalization of the Guruswami-Sudan type decoders. For M-dimensional, or two-dimensional, Reed-Solomon product codes, we are able to show that if the fraction of the number of errors is smaller than $1 - {}^{M(M+1)}\!\!\sqrt{M^M R_p}$, or $1 - \sqrt[6]{4 R_p}$ for $M = 2$, where R_p is the rate of the product code, then the algorithm can efficiently recover the transmitted codeword. The other algorithm is based on the fact that Reed-Solomon product codes can be viewed as subfield-subcodes of Reed-Muller codes. So, the decoding algorithms for Reed-Solomon codes are inherited to decoding of RS product codes. Using the Pellikaan-Wu interpretation for decoding Reed-Muller codes as subcodes of generalized Reed-Solomon codes we prove that if the fraction of the number of errors is smaller than $1 - {}^{2M}\!\!\sqrt{M^M R_p}$, or $1 - \sqrt[4]{4 R_p}$ for $M = 2$, then the algorithm is able to recover the transmitted codeword. For further research directions, it was worth investigating whether product Reed-Solomon codes are subfield subcodes of generalized Reed-Solomon codes with the same length and the same minimum distance. If true one can have a list-decoding algorithm with a radius exceeding half-the-minimum distance of the product code for all rates.

Chapter 8

Performance of Sphere Decoding of Linear Block Codes

When you aim for perfection, you discover it's a moving target.

—George Fisher

Maximum-likelihood (ML) decoding of linear block codes is known to be NP hard [10]. A decoder that utilizes the soft output from the channel directly is called a *soft-decision* (SD) decoder. On the other hand, if hard decisions are made on the received bits before decoding, then such a decoder is called a *hard-decision* (HD) decoder. The optimum decoder is the corresponding HD or SD maximum-likelihood (ML) decoder. Berlekamp's tangential bound is a tighter than the union bound for additive white Gaussian noise (AWGN) channels [11]. Poltyrev derived tight upper bounds on the performance of maximum-likelihood decoding of linear block codes over AWGN channels and binary symmetric (BSC) channels. Bounds based on typical pairs decoding were derived by Aji *et al.* [4]. Other bounds such as the Divsalar simple bound and the variations on the Gallager bounds are tight for AWGN and fading channels [24, 99]. For a broad survey on bounds on the maximum-likelihood decoding of linear codes, see [97].

Fincke and Pohst (FP) [40] described a sphere decoder algorithm which finds the closest lattice point without actually searching all the lattice points. A fast variation of it was given by Schnorr and Euchner [100]. Other efficient closest point search algorithms exist (for a survey see [2]). The sphere decoder algorithm was proposed for decoding lattice codes [113] and for detection in multiple antenna wireless systems [21, 22]. Vikalo and Hassibi proposed HD and SD sphere decoders for joint detection and decoding of linear block codes [110] [111]. On the other hand, one can think of a sphere decoder in a broader sense as any algorithm that returns the closest lattice point to the received word if it exists within a predetermined search radius. By this definition of a sphere decoder, the Berlekamp-Massey algorithm can be considered as a sphere decoder for Reed-Solomon (RS) codes with a search radius equal to half-the-minimum distance of the code. Similarly, the algorithm recently proposed by Guruswami and Sudan for decoding RS codes is an algebraic sphere decoder whose search radius can be larger than half-the-minimum distance of the code [49].

There has a been significant amount of research dedicated to the design of sphere decoders with smaller complexities, complexity analysis of sphere decoders and the application of sphere decoders to various settings and communication systems. However, little research focused on the performance analysis of sphere decoders. This chapter sets down a framework for the analysis of the performance of sphere decoding of block codes over a variety of channels with various modulation schemes.

In this chapter, we study the performance of soft-decision sphere decoding of linear block codes on channels with additive white Gaussian noise and various modulation schemes as BPSK, M-PSK and QAM [89]. This is done in Section 8.1 and Section 8.2 respectively. Bounds on the performance of hard decision sphere decoding on binary symmetric channels (BSC) are derived in Section 8.3. The application of these bounds to the binary image of Reed-Solomon codes is also investigated. We then, in Section 8.4 derive bounds on the maximum-likelihood performance of q-ary linear codes, such as Reed-Solomon codes, over q-ary symmetric channels. This bound becomes handy when analyzing the performance of sphere decoding of Reed-Solomon codes on q-ary symmetric channels. Furthermore, we show, in Section 8.2, how one can analyze the performance of a soft-decision sphere decoder of a general block code with a general modulation scheme. In many settings, we support our analytic bounds by comparing them to numerical simulations. The tradeoff between performance and complexity is discussed in Section 8.5. Finally, we conclude our work in Section 8.6.

8.1 Soft-Decision Sphere Decoding of BPSK and M-PSK Modulated Block Codes

In this section, we consider a sphere decoder when the modulation is binary or M-ary·
phase shift keying (PSK) [89]. Each transmitted codeword in the code has the same
energy when mapped to the PSK constellation. For the case of MPSK modulation,
complex sphere decoding algorithms which solve the closest point search problem were
developed in [58].

8.1.1 Preliminaries

We will introduce some notation, so the bounds derived here are readily applicable for
both M-ary and binary phase shift keying (PSK) modulation. We assume that $\mathcal{C}$ is an
(n, k) linear code. Each codeword of length n will be mapped to a word of M-PSK
symbols. The number of channel symbols will be denoted by n_c. If the code $\mathcal{C}$ is binary
and of length n, then $n_c = \lceil n/\log_2(M) \rceil$. For BPSK, $n_c = n$. Note that the original
code need not be binary. For example, an Reed-Solomon (RS) code defined over $\mathbb{F}_{2^m}$
could be mapped directly to an 2^m-ary PSK constellation by a one-to-one mapping
from the symbols in $\mathbb{F}_{2^m}$ to the 2^m points in the PSK constellation.

For PSK signaling, the code will have the property that all codewords are of equal
energy and lie on a sphere of radius $\sqrt{n_c}$ from the origin of space. Let n_d denote
the dimension of the considered space (noise). For the case of BPSK modulation, the
dimension of the Hamming space is the same as the number of channel symbols (bits)
$n_d = n_c$. On the other hand, for MPSK signaling, $M > 2$, each complex channel
symbol has a real and an imaginary component. Thus the noise has $2 n_c$ independent
components and the dimension of the space is $n_d = 2 n_c$.

Assuming that a codeword $\boldsymbol{c} \in \mathcal{C}$ is transmitted over a binary input AWGN channel,
the received word is $\boldsymbol{y} = \boldsymbol{x} + \boldsymbol{z}$, where $\boldsymbol{x} = \mathcal{M}(\boldsymbol{c})$ and $\mathcal{M}(\boldsymbol{c})$ is the mapping of the
codeword $\boldsymbol{c}$ under PSK modulation, i.e., for BPSK modulation $\mathcal{M}(\boldsymbol{c}) \triangleq 1 - 2\boldsymbol{c}$. The
additive white Gaussian noise (AWGN) is denoted by $\boldsymbol{z} = [z_i]_{i=1}^{n_d}$ with variance σ^2. Let
$E(w)$ be the number of codewords which (after mapping) are at an Euclidian distance
δ_w from each other. Note that for the case of BPSK modulation and a binary code

$\mathcal{C}$, the space is a Hamming space and the Euclidean distance is directly related to the Hamming distance, $\delta_w = 2\sqrt{w}$, where w is the Hamming distance. QPSK modulation and Gray encoding also result in a Hamming space [89] by $\delta_w = \sqrt{2w}$, where w is the (binary) Hamming distance between the codewords. For simplicity in the following analysis, we will assume that the modulated code is linear and the space is a Hamming space.

8.1.2   Analysis of Soft-Decision Sphere Decoding

A soft-decision sphere decoder with an Euclidean radius D, denoted by SSD(D), solves the following optimization problem,

$$\hat{c} = \arg \min_{\boldsymbol{v} \in \mathcal{C}} \quad \|\boldsymbol{y} - \mathcal{M}(\boldsymbol{v})\|^2 \tag{8.1}$$
$$\text{subject to} \quad \|\boldsymbol{y} - \mathcal{M}(\boldsymbol{v})\|^2 \leq D^2,$$

where $\|\boldsymbol{x}\|$ is the Euclidean norm of $\boldsymbol{x}$. Such decoders include *list decoders* that list all codewords whose modulated image is within an Euclidean distance D from the received vector y and choose the closest one. If no such codeword exists, a decoding *failure* is signaled. A decoding *error* is signaled if the decoded codeword is not the transmitted codeword.

Let $\mathcal{E}_D$ denote the event of error or failure of SSD(D), then the error plus failure probability, $P(\mathcal{E}_D)$ is [1]

$$P(\mathcal{E}_D) = P(\mathcal{E}_D | \mathcal{E}_{ML}) P(\mathcal{E}_{ML}) + P(\mathcal{E}_D | \mathcal{S}_{ML}) P(\mathcal{S}_{ML}), \tag{8.2}$$

where $\mathcal{E}_{ML}$ and $\mathcal{S}_{ML}$ denote the events of an ML error and an ML success respectively. Let $\epsilon = \|\boldsymbol{y} - \mathcal{M}(\boldsymbol{c})\|$, then an ML error results if there exists another codeword $\hat{\boldsymbol{c}} \in \mathcal{C}$ such that $\|\boldsymbol{y} - \mathcal{M}(\hat{\boldsymbol{c}})\| \leq \epsilon$. Since limiting the decoding radius to D will not do better than ML decoding, then $P(\mathcal{E}_D | \mathcal{E}_{ML}) = 1$. By observing that $P(\mathcal{S}_{ML}) \leq 1$, it follows

---

[1] Through out this chapter, $P(X)$ will denote the probability that the event X occurs.

that an upper bound on the decoding performance is

$$P(\mathcal{E}_D) \le P(\mathcal{E}_{ML}) + P(\mathcal{E}_D|\mathcal{S}_{ML}). \tag{8.3}$$

Let Ω_D be the Euclidean sphere of radius D centered around the transmitted codeword
in the n_d-dimensional space. The probability that the added white Gaussian noise will
not lie in the sphere Ω_D is

$$P(\boldsymbol{z} \notin \Omega_D) = P\left(\chi_{n_d} > D^2\right) = 1 - \Gamma_r(n_d/2, D^2/2\sigma^2), \tag{8.4}$$

where $\chi_n = \sum_{i=1}^{n} z_i^2$ is a Chi-squared distributed random variable with n degrees
of freedom. Let $\Gamma(x)$ denote the Gamma function, then the cumulative distribution
function (CDF) of χ_v is given by the regularized Gamma function Γ_r [114],

$$\Gamma_r(v/2, w/2) = \begin{cases} \int_0^w \frac{t^{v/2-1}e^{-t/2}}{2^{v/2}\Gamma(v/2)}dt, & w \ge 0 \\ 0, & w < 0 \end{cases}. \tag{8.5}$$

Lemma 8.1. *A lower bound on $P(\mathcal{E}_D)$ is $P(\mathcal{E}_D) \ge P(\boldsymbol{z} \notin \Omega_D)$.*

Proof. The sphere decoder error plus failure probability could be written as

$$\begin{aligned}
P(\mathcal{E}_D) &= P(\mathcal{E}_D|\boldsymbol{z} \in \Omega_D)P(\boldsymbol{z} \in \Omega_D) + P(\mathcal{E}_D|\boldsymbol{z} \notin \Omega_D)P(\boldsymbol{z} \notin \Omega_D) \\
&\ge P(\mathcal{E}_D|\boldsymbol{z} \notin \Omega_D)P(\boldsymbol{z} \notin \Omega_D) \\
&= P(\boldsymbol{z} \notin \Omega_D),
\end{aligned}$$

where the last inequality is because $P(\mathcal{E}_D|\boldsymbol{z} \notin \Omega_D) = 1$ which follows from the definition
of the sphere decoder (8.1). $\square$

Define $\bar{P}(\mathcal{E}_{ML})$ to be an upper bound on the SD-ML decoder error probability, then
we have the following lemma,

Lemma 8.2. $P(\mathcal{E}_D) \le \bar{P}(\mathcal{E}_{ML}) + P(\boldsymbol{z} \notin \Omega_D).$

Proof. Following the proof in the previous lemma,

$$
\begin{aligned}
P(\mathcal{E}_D) &= P(\mathcal{E}_D|\mathbf{z} \in \Omega_D)P(\mathbf{z} \in \Omega_D) + P(\mathcal{E}_D|\mathbf{z} \notin \Omega_D)P(\mathbf{z} \notin \Omega_D) \\
&= P(\mathcal{E}_{ML}, \mathbf{z} \in \Omega_D) + P(\mathcal{E}_D|\mathbf{z} \notin \Omega_D)P(\mathbf{z} \notin \Omega_D) \\
&\leq P(\mathcal{E}_{ML}) + P(\mathbf{z} \notin \Omega_D) \\
&\leq \bar{P}(\mathcal{E}_{ML}) + P(\mathbf{z} \notin \Omega_D).
\end{aligned}
$$

where by definition, $P(\mathcal{E}_{ML}) \leq \bar{P}(\mathcal{E}_{ML})$. $\qquad\square$

Lemma 8.2 provides a way to bound the performance of sphere decoding of linear block codes on a variety of channels where additive white Gaussian noise is added and for a variety of modulation schemes. For example, it can be used in conjunction with the Divsalar bound [24] to give an upper bound on the performance of sphere decoding of linear block codes over independent Rayleigh fading channels. If $\bar{P}(\mathcal{E}_{ML})$ is the union upper bound on the codeword error probability [89, Chapter 8] for BPSK modulation on an AWGN channel, then

$$
P(\mathcal{E}_D) \leq \sum_{w \geq 1} E(w)Q(\sqrt{2\gamma R w}) + P(\mathbf{z} \notin \Omega_D), \tag{8.6}
$$

where $E(w)$ is the number of codewords with (binary) Hamming weight w, γ is the bit signal-to-noise ratio (SNR) and R is the rate of the code.

Lemma 1 implies that one could obtain a tighter upper bound on $P(\mathcal{E}_D)$ by tightening the bound on the ML error probability, $\bar{P}(\mathcal{E}_{ML})$.

8.1.3 The Tangential Sphere Bound

Next, we describe one of the tightest bounds on the soft-decision maximum-likelihood error probability of binary linear codes on binary input AWGN channels, the Poltyrev tangential sphere bound. It is somehow related to Shannon's sphere packing bound [101] which is a lower bound on the error probability where Shannon showed that the Voronoi region of a codeword can be bounded by a right circular n_d-dimensional cone with the codeword on its axis. Poltyrev's tangential sphere bound (TSB) is one of the

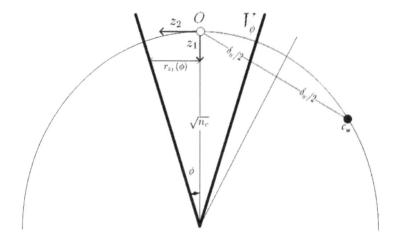

Figure 8.1: Tangential sphere bound: The cone V_ϕ is centered around the transmitted codeword. All codewords lie on a sphere of radius $\sqrt{n_c}$.

tightest bounds on the ML performance of soft-decision decoding of linear codes on AWGN channels with BPSK or MPSK modulation [87, 56] and is calculated by,

$$P(\mathcal{E}_{ML}) \le \min_\theta \left\{ P(\mathcal{E}_{ML}, \mathbf{z} \in V_\theta) + P(\mathbf{z} \notin V_\theta) \right\}, \tag{8.7}$$

where V_θ is an n_d-dimensional right circular cone with a half angle θ whose central line passes through the transmitted codeword and whose apex is at an Euclidean distance $\sqrt{n_c}$ from the transmitted codeword (see Figure 8.1). Let the minimum of the optimization problem in (8.7) be achieved at $\theta = \phi$. For the TSB, the optimum angle ϕ is related to the radius $\sqrt{r_\phi}$ (see Figure 8.2 or Figure 8.3) by $\tan(\phi) = \sqrt{r_\phi/n_c}$, such that r_ϕ is the root of this equation [56]

$$\sum_{\delta_b > 0} E_b'(r_o) \int_0^{\theta_b(r_o)} \sin^{n_d-3}(\vartheta) d\vartheta = \frac{\sqrt{\pi} \Gamma(\frac{n_d-2}{2})}{\Gamma(\frac{n_d-1}{2})} \tag{8.8}$$

when solved for r_o, where $\theta_b(r_o) \triangleq \cos^{-1}\left(\frac{\delta_b/2}{\sqrt{r_o(1-\delta_b^2/4n_c)}}\right)$ and

$$E_b'(r_o) = \begin{cases} E(b), & \delta_b^2/4 < r_o(1-\delta_b^2/4n_c) \\ 0, & \text{otherwise} \end{cases}. \tag{8.9}$$

Let z_1 be the component of the noise along the central axis of the cone with a probability distribution function (PDF) $\mathcal{N}(z_1) = \frac{1}{\sqrt{2\pi\sigma^2}}e^{-z_1^2/2\sigma^2}$ and z_2 be the noise component orthogonal to z_1. Define $\beta_{z_1}(w) \triangleq \frac{\sqrt{n_c}-z_1}{\sqrt{\frac{4n_c}{\delta_w^2}-1}}$ and $r_{z_1}(\phi) \triangleq \sqrt{r_\phi}\left(1 - \frac{z_1}{\sqrt{n_c}}\right)$, then the ML error probability given that the noise z is in the cone V_ϕ is [87]

$$P(\mathcal{E}_{ML}, z \in V_\phi) = \int_{-\infty}^{\infty} \mathcal{N}(z_1)$$
$$\left[\sum_{\delta_b>0} E_b'(r_\phi) \int_{\beta_{z_1}(b)}^{r_{z_1}(\phi)} \mathcal{N}(z_2)\Gamma_r\left(\frac{n_d-2}{2}, \frac{r_{z_1}^2(\phi)-z_2^2}{2\sigma^2}\right) dz_2\right] dz_1. \tag{8.10}$$

8.1.4   A Tight Upper Bound

By Lemma 8.2 and (8.7), we have the following upper bound (which is tighter than (8.6) in case of BPSK)

$$P(\mathcal{E}_D) \leq P(\mathcal{E}_{ML}, z \in V_\phi) + P(z \notin V_\phi) + P(z \notin \Omega_D). \tag{8.11}$$

We observe that instead of directly substituting the TSB of (8.7) for $\bar{P}(\mathcal{E}_{ML})$ in Lemma 8.2 as we did in (8.11), one can find an upper bound which is tighter than (8.11) by noticing that the events $\{z \notin V_\theta\}$ and $\{z \notin \Omega_D\}$ are not, in general, mutually exclusive.

Lemma 8.3. $P(\mathcal{E}_D)$ *is upper bounded by*

$$P(\mathcal{E}_D) \leq P(\mathcal{E}_{ML}, z \in V_\phi) + P(z \notin \Omega_D) + P(\{z \notin V_\phi\} \cap \{z \in \Omega_D\}).$$

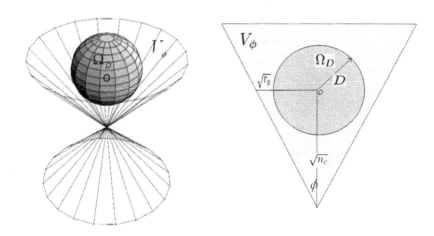

Figure 8.2: Theorem 8.4, *Case A:* The sphere Ω_D lies totally inside the cone V_ϕ ($D \leq \sqrt{n_c}\sin(\phi)$).

Proof. Using Bayes' rule and defining the region $\Lambda(\theta, D) \triangleq \{V_\theta \cap \Omega_D\}$ we get

$$P(\mathcal{E}_D) \leq \min_\theta \{ P(\mathcal{E}_D | \mathbf{z} \in \Lambda(\theta, D)) P(\mathbf{z} \in \Lambda(\theta, D))$$
$$+ P(\mathcal{E}_D | \mathbf{z} \notin \Lambda(\theta, D)) P(\mathbf{z} \notin \Lambda(\theta, D)) \}. \tag{8.12}$$

From the definition of $\Lambda(\theta, D)$, it follows that

$$P(\mathcal{E}_D, \mathbf{z} \in \Lambda(\theta, D)) = P(\mathcal{E}_{ML}, \mathbf{z} \in \Lambda(\theta, D)) \leq P(\mathcal{E}_{ML}, \mathbf{z} \in V_\theta),$$

where the last inequality follows from $\Lambda(\theta, D) \subseteq V_\theta$. Using $P(\mathcal{E}_D | \mathbf{z} \notin \Lambda(\theta, D)) \leq 1$, it follows that

$$P(\mathcal{E}_D) \leq \min_\theta \{ P(\mathcal{E}_{ML}, \mathbf{z} \in V_\theta) + P(\mathbf{z} \notin \Lambda(\theta, D)) \}$$
$$\leq P(\mathcal{E}_{ML}, \mathbf{z} \in V_\phi) + P(\mathbf{z} \notin \{V_\phi \cap \Omega_D\}). \tag{8.13}$$

The last inequality is due to the observation that ϕ does not necessarily minimize (8.13).

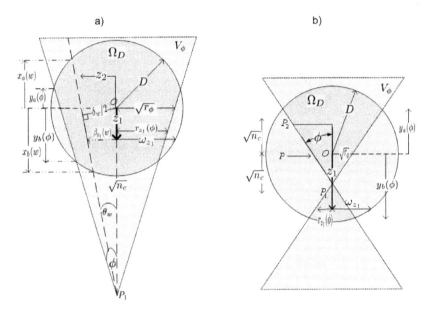

Figure 8.3: Theorem 8.4, *Case B:* The sphere Ω_D intersects the cone V_ϕ; (a) the apex of the cone V_ϕ lies outside the sphere Ω_D ($\sqrt{n_c}\sin(\phi) < D < \sqrt{n_c}$). In case $D \geq \sqrt{n_c}$ (b), the apex of the cone V_ϕ lies inside the sphere Ω_D.

By de Morgan's law, $\{V_\phi \cap \Omega_D\}^c = \{\Omega_D\}^c \cup \{\{V_\phi\}^c \cap \Omega_D\}$, $\{.\}^c$ is the complement of $\{.\}$.   □

We consider two cases;

Case A: The sphere Ω_D lies totally inside the cone V_ϕ. (Figure 8.2). This case is equivalent to the event $\mathbb{A} \triangleq \{D \leq D_\phi\}$, where

$$D_\phi = \sqrt{n_c}\sin(\phi), \qquad (8.14)$$

and will be called the critical decoding radius. It follows that

$$P\left(\{\boldsymbol{z} \notin V_\phi\} \cap \{\boldsymbol{z} \in \Omega_D\}|\mathbb{A}\right) = 0,$$

which could be substituted in Lemma 8.3. Furthermore, since $\Lambda(\theta, D) = \Omega_D$, it follows from (8.12) that a tighter upper bound is

$$P(\mathcal{E}_D|\mathbb{A}) \leq P(\mathcal{E}_{ML}, \mathbf{z} \in \Omega_D) + P(\mathbf{z} \notin \Omega_D). \tag{8.15}$$

The joint probability of the added noise falling inside a sphere of Euclidean radius D and an ML error could be expressed as

$$P(\mathcal{E}_{ML}, \mathbf{z} \in \Omega_D) = \sum_{0 < \frac{\delta_b}{2} < D} E(b) \int_{\frac{\delta_b}{2}}^{D} \mathcal{N}(z_o) \Gamma_r\left(\frac{n_d - 1}{2}, \frac{D^2 - z_o^2}{2\sigma^2}\right) dz_o. \tag{8.16}$$

Let φ be the half angle at which the cone V_φ is tangential to the sphere Ω_D, $\varphi = \sin^{-1}(D/\sqrt{n})$ (see Figure 8.2), then another tight upper bound is

$$P(\mathcal{E}_D|\mathbb{A}) \leq P(\mathcal{E}_{ML}, \mathbf{z} \in V_\varphi) + P(\mathbf{z} \notin \Omega_D). \tag{8.17}$$

Theoretically, it is clear that the bound of (8.15) is tighter than that of (8.17), but numerically they are almost equivalent, since the integration over the region $\{\Omega_D^c \cap V_\varphi\}$ is negligible. Note that $P(\mathcal{E}_{ML}, \mathbf{z} \in V_\varphi)$ is easily calculated using equation (8.10) where $\tan(\varphi) = \sqrt{r_\varphi/n_c}$ and $r_{z_1}(\varphi) = \sqrt{r_\varphi}\left(1 - \frac{z_1}{\sqrt{n_c}}\right)$. $\diamond$

Case B: The sphere Ω_D intersects the cone V_ϕ. (see Figure 8.3). We have two cases depending on the position of the apex of the cone. The first is when the apex of the cone does not lie in the sphere, $\sqrt{n_c}\sin(\phi) < D < \sqrt{n_c}$ (see Figure 8.3a) and the second is when the apex lies in the sphere, $D \geq \sqrt{n_c}$ (see Figure 8.3b). In both cases the following analysis holds. Let the origin, O, of the n_d-dimensional space be at the transmitted codeword which is also the center of Ω_D. Since the cone and the sphere are symmetrical around the central axis, we project on a two-dimensional plane as in Figure 8.3. The radial component of the noise (along the axis of the cone) is z_1. The altitudes $y_a(\phi)$ and $y_b(\phi)$ at which the (double) cone intersects the sphere are found by substituting the line equation $P = P_1 + U(P_2 - P1)$, where $P = (x, y)$, $P_1 = (0, \sqrt{n_c})$ and $P_2 = (2\sqrt{n_c}\tan(\phi), -\sqrt{n_c})$ into the quadratic equation of the sphere. It follows that $y_{a,b}(\phi) = \sqrt{n_c}(1 - 2U_{a,b}(\phi, D))$, where

$$U_{a,b}(\theta, D) = \frac{4n_c \pm \sqrt{16n_c^2 - 16n_c \sec^2(\theta)(n_c - D^2)}}{8n_c \sec^2(\theta)}.$$

It is easy to check that at $D = \sqrt{n_c}$, $u_b = 0$ and y_b is at the apex of V_ϕ. If $D > \sqrt{n_c}$ then the intersection at y_b is in the lower nappe of the cone. It is also observed that V_ϕ and Ω_D do not intersect ($\Omega_D \subset V_\phi$) if $16n_c^2 < 16n_c \sec^2(\phi)(n_c - D^2)$ or equivalently $D < \sqrt{n_c}\sin(\phi)$ which is Case A.

Define $\mathbb{B}$ to be the event $\mathbb{B} \triangleq \{D > \sqrt{n_c}\sin(\phi)\}$, $f_{n-1}(t)$ to be the PDF of $\chi_{n-1} = \sum_{i=2}^{n} z_i^2$, and $w_{z_1}^2 = D^2 - z_1^2$ (see Figure 8.3). From Lemma 8.3, the error probability is upper bounded by

$$P(\mathcal{E}_D|\mathbb{B}) \le P(\mathcal{E}_{ML}, \mathbf{z} \in V_\phi) + P(\mathbf{z} \notin \Omega_D) + P(\{\mathbf{z} \notin V_\phi\} \cap \{\mathbf{z} \in \Omega_D\}|\mathbb{B}), \quad (8.18)$$

and

$$P(\{\mathbf{z} \notin V_\phi\} \cap \{\mathbf{z} \in \Omega_D\}|\mathbb{B}) = \int_{y_a(\phi)}^{y_b(\phi)} \mathcal{N}(z_1) \int_{r_{z_1}^2(\phi)}^{w_{z_1}^2} f_{n_d-1}(t)\mathrm{d}t\mathrm{d}z_1 \quad (8.19)$$

by Figure 8.3.   ◇

The tight upper bound is summarized in this theorem,

Theorem 8.4. *The performance of soft-decision sphere decoding with an Euclidean decoding radius D of a linear code with (Euclidean) weight spectrum $E(b)$ on an AWGN channel with noise variance σ^2 and (binary or M-ary) PSK modulation is upper bounded by:*

$$P(\mathcal{E}_D) \le \begin{cases} \sum_{0 < \frac{\delta_b}{2} < D} E(b) \int_{\frac{\delta_b}{2}}^{D} \frac{e^{-z_o^2/2\sigma^2}}{\sqrt{2\pi\sigma^2}} \Gamma_r\left(\frac{n_d-1}{2}, \frac{D^2-z_o^2}{2\sigma^2}\right) \mathrm{d}z_o \\ \quad +1 - \Gamma_r(n_d/2, D^2/2\sigma^2), \hfill D \le \sqrt{n_c}\sin(\phi) \\[2ex] \int_{-\infty}^{\infty} \mathcal{N}(z_1) \sum_{\delta_b > 0} E_b'(r_\phi) \\ \quad \int_{\beta_{z_1}(b)}^{r_{z_1}(\phi)} \mathcal{N}(z_2) \Gamma_r\left(\frac{n_d-2}{2}, \frac{r_{z_1}^2(\phi)-z_2^2}{2\sigma^2}\right) \mathrm{d}z_2\mathrm{d}z_1 \\ \quad +1 - \Gamma_r(n_d/2, D^2/2\sigma^2) \\ \quad + \int_{y_a(\phi)}^{y_b(\phi)} \left(\Gamma_r\left(\frac{n_d-1}{2}, \frac{w_{z_1}^2}{2\sigma^2}\right) - \Gamma_r\left(\frac{n_d-1}{2}, \frac{r_{z_1}^2(\phi)}{2\sigma^2}\right)\right) \mathcal{N}(z_1)\mathrm{d}z_1, \quad D > \sqrt{n_c}\sin(\phi) \end{cases}$$

where ϕ is the half angle of the cone V_ϕ and is given by (8.8).

Following the proof of Lemma 8.3, the error plus failure probability of SSD(D) is upper bounded by

$$P(\mathcal{E}_D) \leq P(\mathcal{E}_D, \mathbf{z} \in \Lambda(\phi, D)) + P(\mathbf{z} \notin \Lambda(\phi, D)). \tag{8.20}$$

From the previous arguments in *Case A* and *Case B*, the following theorem provides a slightly tighter upper bound than that of the previous theorem.

Theorem 8.5. *The performance of SSD(D) for BPSK or MPSK modulation is upper bounded by*

$$P(\mathcal{E}_D) \leq \begin{cases} P(\mathcal{E}_{ML}, \mathbf{z} \in \Omega_D) + P(\mathbf{z} \notin \Omega_D), & D \leq \sqrt{n_c}\sin(\phi) \\ P(\mathcal{E}_{ML}, \mathbf{z} \in \Lambda(\phi, D)) + P(\mathbf{z} \notin \Omega_D) \\ \quad + P(\{\mathbf{z} \notin V_\phi\} \cap \{\mathbf{z} \in \Omega_D\}), & D > \sqrt{n_c}\sin(\phi) \end{cases}.$$

Observe that the difference from Theorem 8.4 is that the term $P(\mathcal{E}_{ML}, \mathbf{z} \in \Lambda(\phi, D))$ was upper bounded by $P(\mathcal{E}_{ML}, \mathbf{z} \in V(\phi))$ in Theorem 8.4. Consider a codeword at a distance δ_w, then the half angle of the cone bisecting this distance is $\theta_w = \sin^{-1}(\delta_w/2\sqrt{n_c})$ (Figure 8.3). This cone will intersect the sphere Ω_D at altitudes $x_a(w)$ and $x_b(w)$ given by $x_{a,b}(w) = \sqrt{n_c}(1 - 2U_{a,b}(\theta_w, D))$. Now define the integral

$$\mathcal{I}_2(w) = \int_{x_a(w)}^{y_a(\phi)} \mathcal{I}(\omega_{z_1}, w, z_1)\mathrm{d}z_1 + $$
$$\int_{y_a(\phi)}^{y_b(\phi)} \mathcal{I}(r_{z_1}(\phi), w, z_1)\mathrm{d}z_1 + \int_{y_b(\phi)}^{x_b(w)} \mathcal{I}(\omega_{z_1}, w, z_1)\mathrm{d}z_1, \tag{8.21}$$

where

$$\mathcal{I}(\gamma, w, z_1) \triangleq \mathcal{N}(z_1) \int_{\beta_{z_1}(w)}^{\gamma} \mathcal{N}(z_2)\Gamma_r\left(\frac{n_d - 2}{2}, \frac{\gamma^2 - z_2^2}{2\sigma^2}\right)\mathrm{d}z_2. \tag{8.22}$$

Taking the union over all codewords with nonzero Euclidean weights such that $\theta_w < \phi$, it follows that for $D > \sqrt{n_c}\sin(\phi)$,

$$P(\mathcal{E}_{ML}, \mathbf{z} \in \Lambda(\phi, D)) = \sum_{\delta_b > 0} E_b'(r_\phi)\mathcal{I}_2(w), \tag{8.23}$$

and $E_b'(r_\phi)$ is given by (8.9). It is to be noted that the same equations hold whether $(D \geq \sqrt{n_c})$ or $(\sqrt{n_c}\sin(\phi) < D < \sqrt{n_c})$.

8.1.5   A Note on Reed-Solomon Codes

Consider the case when the binary image of an Reed-Solomon (RS) code, defined over $\mathbb{F}_{2^m}$, is transmitted over an AWGN channel and the decoder is either a HD or SD sphere decoder. Tight upper bounds on the performance of hard-decision and soft-decision maximum likelihood decoding of the binary images of Reed-Solomon codes were developed in Section 2.4 by averaging over all possible binary representations of the RS code. We will use the same technique in this chapter to analyze the performance of the sphere decoders when the code of interest is the binary image of an RS code. In this case the average binary weight enumerator of the ensemble of binary images of an RS code will be used as the weight enumerator in this analysis.

8.1.6   Numerical Results

In Figure 8.4, we show how the bounds derived for M-ary modulated spherical codes are tight. The simulation curves and the analytical bounds will be labeled by "sim" and "bnd" respectively. A codeword in the $(24, 12)$ Golay code is mapped into 12 QPSK symbols and transmitted over an AWGN channel. As observed, the simulated performance of the ML decoder and the SD sphere decoder [110] are tightly bounded by the bounds given in this section. The critical decoding radius in the 2×12 -dimensional space is $D_\phi = 2.667$.

In Figure 8.5, the performance of SD sphere decoding of the binary image of the $(15, 11)$ RS code, BPSK modulated over an AWGN channel, is investigated. The ML performance is simulated by means of the MAP decoder, and it is observed that the averaged ML bound is tight [29]. We simulated the performance of SD sphere decoding when the decoding radii were 3 and 3.5 respectively. Our analytical bounds almost overlapped with the simulations. The critical decoding radius is $D_\phi = 4.588$. A decoder with an Euclidean decoding radius of 5 has a near ML performance at an SNR of 5 dB. For reference purposes, we plot the performance of the hard-decision

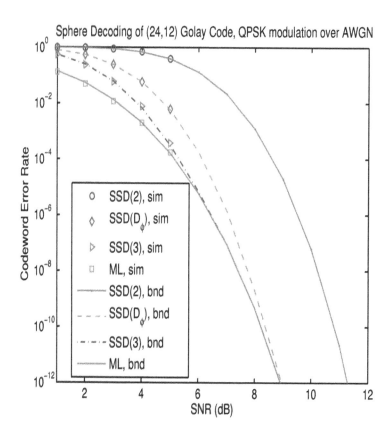

Figure 8.4: Bounds on the performance of soft-decision sphere decoding of the (24, 12) Golay code when QPSK modulated over an AWGN channel.

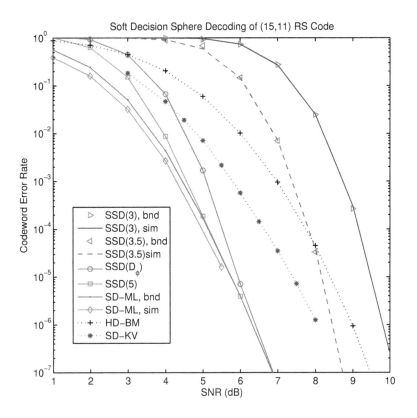

Figure 8.5:   Bounds on the performance of SSD of a binary image of the $(15, 11)$ Reed-Solomon code BPSK modulated on an AWGN channel.

Berlekamp-Massey (BM) decoder and the algebraic soft-decision decoder by Koetter and Vardy [72]. It is worth noting that algebraic soft decoding can also achieve near ML performance [31, 33].

8.2 Sphere Decoding of Lattices

In this section, we consider the case of soft-decision sphere decoding of a general lattice or code $\mathcal{C}$. In contrast to the case of Section 8.1 the code is not constrained to be a linear code and the transmitted codewords are not constrained to have a fixed energy. The channel symbols of a transmitted codeword are also not required to have the same energy.

Define $E(i, w)$ to be the number of mapped codewords with an Euclidean distance δ_w from the ith codeword. Given that $\mathbf{c}_i$ is transmitted, let the error probability of SSD(D) be upper bounded by $P_i(\mathcal{E}_D)$. By taking the expectation over all codewords,

$$P(\mathcal{E}_D) \leq \frac{1}{|\mathcal{C}|} \sum_{\mathbf{c}_i \in \mathcal{C}} P_i(\mathcal{E}_D). \tag{8.24}$$

Now, if we assume that $P_i(\mathcal{E}_D)$ is of the union bound form;

$$P_i(\mathcal{E}_D) = \sum_w E(i, w) P_i^{(w)}(\mathcal{E}_D),$$

where $P_i^{(w)}(\mathcal{E}_D)$ is the probability of a sphere decoder error due to incorrectly decoding a codeword at a distance δ_w when $\mathbf{c}_i$ is transmitted. The error probability of SSD(D) can thus be upper bounded by

$$P(\mathcal{E}_D) \leq \sum_{\delta_w > 0} \bar{E}(w) P^{(w)}(\mathcal{E}_D),$$

where $P^{(w)}(\mathcal{E}_D)$ is the probability that the sphere decoder erroneously decodes a codeword at a distance δ_w from the transmitted codeword and

$$\bar{E}(w) = \frac{1}{|\mathcal{C}|} \sum_{c_i \in \mathcal{C}} E(i, w), \tag{8.25}$$

is the average number of codewords which are at an Euclidean distance δ_w from another codeword. For an arbitrary finite code or lattice $\mathcal{C}$, using arguments from the previous sections, the error probability $SSD(D)$ can be upper bounded by

$$P(\mathcal{E}_D) \leq \min_{D' \leq D} \{P(\mathcal{E}_{ML}, \mathbf{z} \in \Omega_{D'}) + P(\mathbf{z} \notin \Omega_{D'})\}, \tag{8.26}$$

where $P(\mathbf{z} \notin \Omega_D)$ is given by (8.4) and

$$P(\mathcal{E}_{ML}, \mathbf{z} \in \Omega_D) = \sum_{0 < \frac{\delta_w}{2} < D} \bar{E}(w) \int_{\frac{\delta_w}{2}}^{D} \frac{1}{\sqrt{2\pi\sigma^2}} e^{-z^2/2\sigma^2} \Gamma_r\left(\frac{n_d - 1}{2}, \frac{D^2 - z^2}{2\sigma^2}\right) dz. \tag{8.27}$$

The Hughes upper bound on the ML error probability is $P(\mathcal{E}_{ML}) \leq \min_D P(\Psi(D))$ [61], where

$$\Psi(D) \triangleq P(\mathcal{E}_{ML}, \mathbf{z} \in \Omega_D) + P(\mathbf{z} \notin \Omega_D). \tag{8.28}$$

The radius D_o that minimizes this error probability is the root of the equation [55]

$$\sum_{0 < \frac{\delta_w}{2} < D} \bar{E}(w) \int_0^{\theta_{w,D}} \sin(\theta)^{n_d - 2} d\theta = \frac{\sqrt{\pi} \Gamma\left(\frac{n_d - 1}{2}\right)}{\Gamma\left(\frac{n_d}{2}\right)}, \tag{8.29}$$

where $\theta_{w,d} = \cos^{-1}(\delta_w/2D)$. From (8.26), the upper bound on the sphere decoding error probability is given by

$$P(\mathcal{E}_D) \leq \begin{cases} \Psi(D), & D < D_o \\ \Psi(D_o), & D \geq D_o \end{cases}.$$

Furthermore, the optimum radius D_o does not depend on the channel and can be the radius of choice for near maximum-likelihood decoding. The bound developed here is universal in the sense that also applies for the case of a linear code with equal energy codewords. However, it is to be noted that the Hughes bound on ML decoding is not tighter than the Poltyrev tangential sphere bound [23].

For the case of M-PSK modulation of a linear code, the constellation may not result in a Hamming space if $M > 4$. In such a case the ensemble average weight enumerator $\bar{E}(w)$ can be used with the bounds of Section 8.1 to analyze the performance. (The same technique can also be used with the results in next sections.)

Example 8.1. Assume an $(15,3)$ RS code over $\mathbb{F}_{16}$ and assume a one-to-one mapping from the symbols of $\mathbb{F}_{16}$ to the points of an 16-QAM modulation [89], whose average energy per symbol is 10. The ensemble weight enumerator $\bar{E}(w)$ was numerically computed to evaluate the bounds. The radius that minimizes the bound on the ML error probability is $D_o = 12.9$. In Figure 8.6, we confirm that the bounds on the sphere decoder error probability agree with the simulations for the case of $D = 10$. We also compare the simulated performance of ML error probability $P(\mathcal{E}_{ML}, \boldsymbol{z} \in \Omega_D)$ to that of the analytic performance in both cases. At low SNRs this probability is low as the probability of the received word falling inside the sphere is relatively low. As more received words fall inside the sphere, the ML error probability increases as the SNR increases. At a certain SNR, the probability of the ML error starts decreasing due to the improved reliability of the received word.   ⋄

8.3 Sphere Decoding on Binary Symmetric Channels

In this section, an upper bound on the performance of the hard-decision sphere decoder, when the code is transmitted over the BSC, is derived. Transmitting a binary codeword over a binary input AWGN channel followed by hard decisions is equivalent to transmitting it on a BSC with a crossover probability $p = Q(\sqrt{2R\gamma})$ where γ is the bit signal-to-noise ratio. In case of M-PSK signaling with gray encoding, $p \approx \frac{p_c}{\log_2(M)}$ where $p_c = 2Q\left(\sqrt{2k\gamma}\sin\frac{\pi}{M}\right)$ [89].

Let $\boldsymbol{y}$ be the received word when the codeword $\boldsymbol{c}$ is transmitted over an BSC channel. The HD sphere decoder with radius m, $\mathrm{HSD}(m)$, finds the codeword $\hat{\boldsymbol{c}}$, if it

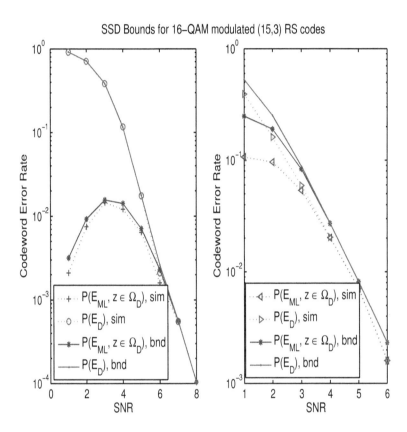

Figure 8.6:   Performance of soft-decision sphere decoding of the $(15,3)$ RS code, 16-QAM modulated, and transmitted over an AWGN channel.
The soft-decision sphere decoders have an Euclidean radius 10 (left) and $D_o = 12.9$ (right). The bounds are compared to simulations for a sphere decoding ML error $P(\mathcal{E}_{ML}, z \in \Omega_D)$ and the error plus failure probability $P(\mathcal{E}_D)$.

exists, such that

$$\hat{c} = \arg\min_{v \in \mathcal{C}} \quad d\,(y, v) \tag{8.30}$$
$$\text{subject to} \quad d(y, v) < m + 1,$$

where $d\,(y, v)$ is the Hamming distance between y and v. Let $\zeta = d(y, c)$ then, from the linearity of the code, the probability that the received word is outside a Hamming sphere (ball) of radius $m - 1$ centered around the transmitted codeword is

$$P(\zeta \geq m) = \sum_{t=m}^{n} \binom{n}{t} p^t (1 - p)^{n-t}. \tag{8.31}$$

Poltyrev [87] derived a tight bound on the performance of the HD-ML decoder based on,

$$P(\mathcal{E}_{ML}) \leq \min_{m} \left\{ P(\mathcal{E}_{ML}, \zeta < m) + P(\zeta \geq m) \right\}. \tag{8.32}$$

The minimum of the above equation is at m_o where m_o is the smallest integer m such that [87]

$$\sum_{b=1}^{2m} E(b) \sum_{r=\lceil \frac{b}{2} \rceil}^{m} \binom{b}{r} \binom{n-b}{m-r} \geq \binom{n}{m}. \tag{8.33}$$

We now turn our attention to the hard-decision sphere decoder with an arbitrary decoding radius. Let $P(\Sigma_m)$, be the error plus failure probability of the hard decision sphere decoder, $\text{HSD}(m - 1)$, then $P(\Sigma_m)$ could be written as

$$P(\Sigma_m) = P(\Sigma_m, \zeta < m) + P(\Sigma_m | \zeta \geq m) P(\zeta \geq m)$$
$$= P(\mathcal{E}_{ML}, \zeta < m) + P(\zeta \geq m), \tag{8.34}$$

where we used the fact that $P(\Sigma_m | \zeta \geq m) = 1$ and the observation that given that $\zeta < m$, the conditional error probability of the $\text{HSD}(m - 1)$ and the HD-ML decoders are the same. The last term in the above equation is a lower bound on the failure probability of the $\text{HSD}(m - 1)$ decoder. The joint probability of an HD-ML error and

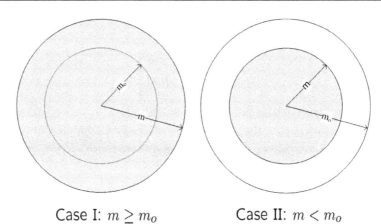

Case I: $m \geq m_o$ Case II: $m < m_o$

Figure 8.7:   Two cases for the bound on the performance of hard-decision sphere decoders (Theorem 8.11).

$d(\boldsymbol{y}, \boldsymbol{c}) < m$ is upper bounded by the union bound [87],

$$
P(\mathcal{E}_{ML}, \zeta < m) \leq \sum_{b=1}^{2(m-1)} E(b) \sum_{r=\lceil \frac{b}{2} \rceil}^{m-1} \left[\binom{b}{r} p^r (1-p)^{b-r} \sum_{s=0}^{m-r-1} \binom{n-b}{s} p^s (1-p)^{n-b-s} \right].
$$
(8.35)

Similar to the soft-decision decoding case, we have the following lemma:

Lemma 8.6. *A lower bound on the performance of a hard decision sphere decoder, HSD($m-1$), over a BSC with parameter p is* $P(\Sigma_m) \geq \sum_{t=m}^{n} \binom{n}{t} p^t (1-p)^{n-t}$.

To develop a tight upper bound on $P(\Sigma_m)$, we consider two cases (see Figure 8.7):
Case I: The decoding radius $m \geq m_o$. Equation (8.34) can be written as

$$
P(\Sigma_m | m \geq m_o) = P(\mathcal{E}_{ML}, \zeta < m_o) + P(\mathcal{E}_{ML}, m_o \leq \zeta < m) + P(\zeta \geq m).
$$

It follows that

$$
P(\Sigma_m | m \geq m_o) \leq P(\mathcal{E}_{ML}, \zeta < m_o) + P(\zeta \geq m_o). \tag{8.36}
$$

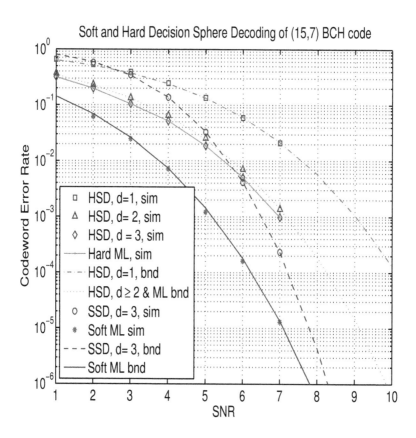

Figure 8.8: Bounds on the codeword error rate of soft-decision and hard-decision sphere decoding of the (15, 7) BCH code BPSK modulated over an AWGN channel. The simulations (labeled by "sim") are tightly upper bounded by the analytic bounds (labeled by "bnd").

We observe that the upper bound reduces to that of the HD-ML case (8.32). By recalling that the minimum of (8.32) is achieved at m_o, the bound of (8.34) is looser than (8.36) when $m > m_o$. The intuition behind this is that the performance of a sphere decoder with a decoding radius $m_o - 1$ or greater approaches that of the ML decoder.

Case II: The decoding radius $m < m_o$. Noticing that the sphere

$$\{\zeta < m\} \subset \{\zeta < m_o\}, \ P(\Sigma_m | m < m_o)$$

is indeed given by (8.34).

Thus, we have proved the following theorem,

Theorem 8.7. *The performance of a hard-decision sphere decoder with a decoding radius $m - 1$ when used for decoding a linear code with a weight spectrum $E(b)$ over an BSC channel with a crossover probability p is upper bounded by*

$$P(\Sigma_m) \leq \begin{cases} P(\mathcal{E}_{ML}, \zeta < m_o) + P(\zeta \geq m_o), & m \geq m_o \\ P(\mathcal{E}_{ML}, \zeta < m) + P(\zeta \geq m), & m < m_o \end{cases}, \quad (8.37)$$

where m_o is radius that minimizes (8.32) and is the solution of (8.33). $P(\zeta \geq m)$ is given by (8.31) and $P(\mathcal{E}_{ML}, \zeta < m)$ is given by (8.35).

8.3.1 Numerical Examples

In this subsection, the bounds developed for SD and HD sphere decoding are evaluated and compared with the performance of the corresponding sphere decoders, [110] and [111] respectively.

In Figure 8.8, we compare the analytical bounds to simulations of sphere decoding of an (15, 7) BCH code BPSK modulated and transmitted over an AWGN channel. The minimum distance of the BCH code is 5. The critical decoding Euclidian radius of the soft-decision decoder is $D_\phi = 3.17$ while the critical Hamming decoding radius of the hard decision decoder is $m_o = 3$. We observe that the simulated performance is tightly upper bounded by the analytical bounds of Theorem 8.4 and Theorem 8.11 for soft and hard decision sphere decoding respectively. The larger the decoding radius the nearer the performance is to maximum-likelihood decoding.

8.4 Sphere Decoding on q-ary Symmetric Channels

Now consider an (n, k, d) RS code and a hard-decision sphere decoder which can correct τ symbol errors, where the symbols are in $\mathbb{F}_q$. The Berlekamp-Massey algorithm is a well-known polynomial time algorithm that can correctly decode words which are at a (symbol) Hamming distance of $\tau_{BM} = \lfloor \frac{n-k}{2} \rfloor$ from the transmitted codeword. The error probability of bounded distance decoding of RS codes is well studied (c.f., [79]). Recently, Guruswami and Sudan [49] developed a list-decoding algorithm that can correct up to $\tau_{GS} = \lceil n - \sqrt{nk} - 1 \rceil$ symbol errors. To analyze this case, we first derive a bound on the performance of the corresponding ML decoder.

8.4.1 Maximum Likelihood Decoding of Linear Block Codes on q-ary Symmetric Channels

We will assume an (n, k, d) linear code over $\mathbb{F}_q$ transmitted over a q-ary symmetric channel. The probability that a symbol is correctly received will be denoted by s, while the probability that it is received as another symbol will be $p = (1 - s)/(q - 1)$. Transmitting a q-ary code over an AWGN channel followed by hard-decision can be modeled as transmitting it over a q-ary symmetric channel. Assume that $q = 2^m$, the channel alphabet size is 2^b, $b \le m$, and each q-ary symbol is mapped to m/b channel symbols. Let p_c be the probability that a channel symbol is incorrectly decoded, then $s = (1 - p_c)^{m/b}$. For example, if the channel is a BPSK channel with a bit signal-to-noise ratio γ, $q = 2^m$ and the binary image of the RS code is transmitted, then a q-ary symbol is correctly received if all the m bits in its binary image are correctly received, i.e., $s = \left(1 - Q\left(\sqrt{2\frac{k}{n}\gamma}\right)\right)^m$.

Let ζ be the Hamming distance between the transmitted codeword and the received q-ary word. Then, similar to the binary case, the ML error probability can be upper bounded as follows,

$$P(\mathcal{E}_{ML}) \leq \min_m \left\{ P(\mathcal{E}_{ML}, \zeta < m) + P(\zeta \geq m) \right\}. \tag{8.38}$$

Assuming that the code is linear, the probability that the received q-ary word lies outside a Hamming sphere (ball) of radius $m - 1$ centered around the transmitted word is

$$P(\zeta \geq m) = \sum_{\alpha=m}^{n} \binom{n}{\alpha} (1 - s)^{\alpha} s^{n-\alpha}. \tag{8.39}$$

The above equation will also provide a lower bound on the performance of the sphere decoder.

The first term in (8.38) is upper bounded in the following lemma.

Lemma 8.8. *For an* (n, k, d) *linear code over* $\mathbb{F}_q$, *with a weight enumerator* $E(w)$, *transmitted over a* q-ary *symmetric channel with parameters* s *and* p,

$$P(\mathcal{E}_{ML}, \zeta < m) \leq \sum_{w=d}^{\min\{n, 2(m-1)\}} E(w) \sum_{\alpha=0}^{\min\{w, m-1\}} \sum_{\eta = \lceil \frac{w-\alpha}{2} \rceil}^{w-\alpha} \tag{8.40}$$

$$\left(\frac{w!}{\eta! \alpha! (w - \eta - \alpha)!} p^{\eta} (1 - p - s)^{\alpha} s^{w-\eta-\alpha} \sum_{\beta=0}^{m-1-\eta-\alpha} \binom{n-w}{\beta} (1 - s)^{\beta} s^{n-w-\beta} \right).$$

Proof. We will assume that the all-zero codeword is transmitted. Now consider a codeword $\boldsymbol{c}$ with Hamming weight w and assume the received word $\boldsymbol{r}$ has a Hamming weight $m' - 1$ (see Figure 8.9). Consider the w nonzero symbols in $\boldsymbol{c}$ and the corresponding coordinates in $\boldsymbol{r}$. Let $\boldsymbol{r}$ and $\boldsymbol{c}$ have the same symbols in η of these coordinates. Let α of these w coordinates in $\boldsymbol{r}$ be neither zero nor match those in $\boldsymbol{c}$, and $w - \eta - \alpha$ of the remaining coordinates be zero. Since the Hamming weight of $\boldsymbol{r}$ is $m' - 1$, there must be $m' - 1 - \eta - \alpha$ nonzero symbols in the remaining $n - w$ coordinates and the remaining symbols will be zero. The probability of receiving such a word is $\frac{w!}{\eta! \alpha! (w-\eta-\alpha)!} p^{\eta} (1 - p - s)^{\alpha} s^{w-\eta-\alpha} \binom{n-w}{m'-1-\eta-\alpha} (1 - s)^{m'-1-\eta-\alpha} s^{n-w-(m'-1-\eta-\alpha)}$. In such a case, the Hamming distance between $\boldsymbol{r}$ and $\boldsymbol{c}$ is $w + m' - 1 - 2\eta - \alpha$. An ML error result if this is less than the weight of $\boldsymbol{r}$, i.e., if $\eta \geq \lceil \frac{w-\alpha}{2} \rceil$. By summing over all

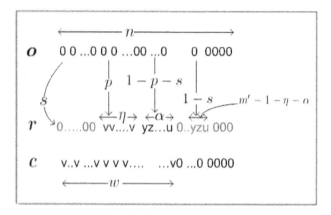

Figure 8.9: Proof of Lemma 8.8.

possible combinations of η and α and applying the union bound for all codewords that can be within a Hamming distance m' from r, the error probability is upper bounded by

$$\sum_{w=d}^{\min\{n,2(m'-1)\}} E(w) \sum_{\alpha=0}^{\min\{w,m'-1\}} \sum_{\eta=\lceil\frac{w-\alpha}{2}\rceil}^{w-\alpha} \left(\frac{w!}{\eta!\alpha!(w-\eta-\alpha)!}p^\eta(1-p-s)^\alpha s^{w-\eta-\alpha}\right.$$
$$\left.\binom{n-w}{m'-1-\eta-\alpha}\left((1-s)^{m'-1-\eta-\alpha}s^{n-w-(m'-1-\eta-\alpha)}\right)\right).$$

Applying the union bound for all received words with Hamming weights less than m, $m' \leq m$, the result follows. □

We are now ready to prove the following theorem,

Theorem 8.9. *The maximum-likelihood error probability of an (n, k, d) q-ary linear* *code on a q-ary symmetric channel is upper bounded by*

$$P(\mathcal{E}_{ML}) \leq \sum_{w=d}^{\min\{n,2(m_o-1)\}} E(w) \sum_{\alpha=0}^{\min\{w,m_o-1\}} \sum_{\eta=\lceil \frac{w-\alpha}{2} \rceil}^{w-\alpha} \left(\frac{w!}{\eta! \alpha! (w-\eta-\alpha)!} p^\eta (1-p-s)^\alpha \right.$$

$$\left. s^{w-\eta-\alpha} \sum_{\beta=0}^{m_o-1-\eta-\alpha} \binom{n-w}{\beta} (1-s)^\beta s^{n-w-\beta} \right) + \sum_{\alpha=m_o}^{n} \binom{n}{\alpha} (1-s)^\alpha s^{n-\alpha},$$

where m_o is the smallest integer m such that

$$\sum_{w=d}^{\min\{n,2m\}} E(w) \sum_{\alpha=0}^{\min\{w,m\}} \left(\frac{q-2}{q-1} \right)^\alpha$$

$$\sum_{\eta=\lceil \frac{w-\alpha}{2} \rceil}^{w-\alpha} \left(\frac{1}{q-1} \right)^\eta \frac{w!}{\eta! \alpha! (w-\eta-\alpha)!} \binom{n-w}{m-\eta-\alpha} \geq \binom{n}{m}. \qquad (8.41)$$

Proof. The upper bound follows by substituting (8.40) and (8.39) in (8.38). Observe that $P(\mathcal{E}_{ML}) \leq P(\mathcal{E}_{ML}, \zeta < m) + P(\zeta \geq m)$ and $P(\mathcal{E}_{ML}, \zeta < m)$ is increasing in m while $P(\zeta \geq m)$ is decreasing in m. By discrete differentiation, the minimum is achieved at m such that

$$(P(\mathcal{E}_{ML}, \zeta < m+1) - P(\mathcal{E}_{ML}, \zeta < m)) \geq (P(\zeta \geq m) - P(\zeta \geq m+1)).$$

Optimizing over the radius m, the minimum is thus achieved at the first integer m such that

$$\sum_{w=d}^{2m} E(w) \sum_{\alpha=0}^{m} \sum_{\eta=\lceil \frac{w-\alpha}{2} \rceil}^{w-\alpha} \left(\frac{w!}{\eta! \alpha! (w-\eta-\alpha)!} p^\eta (1-p-s)^\alpha s^{w-\eta-\alpha} \right.$$

$$\left. \left(\binom{n-w}{m-\eta-\alpha} (1-s)^{m-\eta-\alpha} s^{n-w-m+\eta+\alpha} \right) \geq \binom{n}{m} (1-s)^m s^{n-m},$$

which reduces to the condition of (8.41). $\qquad\qquad \square$

It is worth noting that the optimum radius m_o which minimizes the bound on the ML error probability only depends on the weight enumerator of the code and the size of its finite field. Since the optimum radius does not depend on the SNR, it is valid for q-ary symmetric channels at any SNR. Similar to the binary case [87], we establish below a connection between m_o and the covering radius of the code.

Lemma 8.10. *The covering radius of a linear code on $\mathbb{F}_q$ is lower bounded by $m_o - 1$, where m_o is given by Theorem 8.9.*

Proof. Define $L(m)$ to be the left hand side term in (8.41) and $\boldsymbol{c}_o$ to be the all zero codeword. Similar to the proof of Lemma 8.8, one can show that

$$(q-1)^m L(m) = |\{\boldsymbol{r} \in \mathbb{F}_q^n : \ \mathrm{d}(\boldsymbol{r}, \boldsymbol{c}_o) = m \ \text{ and } \ \mathrm{d}(\boldsymbol{r}, \boldsymbol{c}_i) \leq m \ \text{ for some } \ \boldsymbol{c}_i \in \mathcal{C} \setminus \boldsymbol{c}_o \}|.$$

Also,

$$(q-1)^m \binom{n}{m} = |\{\boldsymbol{r} \in \mathbb{F}_q^n : \mathrm{d}(\boldsymbol{r}, \boldsymbol{c}_o) = m \}|.$$

Since $(q-1)^{m_o-1} L(m_o-1) < (q-1)^{m_o-1} \binom{n}{m_o-1}$, it follows that there exits words $\boldsymbol{r} \in \mathbb{F}_q^n$ such that $\min_{\boldsymbol{c} \in \mathcal{C}} \mathrm{d}(\boldsymbol{r}, \boldsymbol{c}) = m_o - 1$ and this minimum is achieved when $\boldsymbol{c}$ is the all zero codeword $\boldsymbol{c}_o$. By recalling that the covering radius is [74]

$$R_c = \max_{\boldsymbol{r} \in \mathbb{F}_q^n} \min_{\boldsymbol{c} \in \mathcal{C}} \mathrm{d}(\boldsymbol{r}, \boldsymbol{c}),$$

it follows that $R_c \geq m_o - 1$.   □

8.4.2 Hard-Decision Sphere Decoding of Linear Block Codes on q-ary Symmetric Channels

Here, we consider the case when the decoder is a q-ary hard decision sphere decoder. As for the binary case, the HSD$(m-1)$ can correctly decode a codeword if the number of q-ary symbol errors is $m-1$ or less. Thus the error plus failure probability of the q-ary hard decision sphere decoder will be bounded by this theorem.

Theorem 8.11. *The performance of a hard-decision sphere decoder with a decoding radius $m-1$ when used for decoding a linear code with a weight spectrum $E(b)$ over*

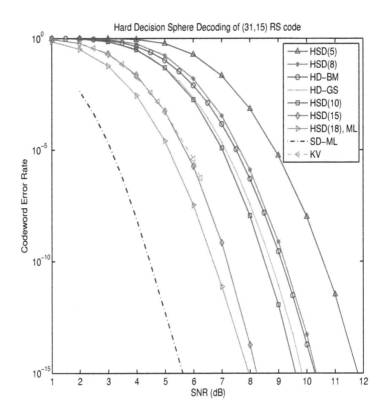

Figure 8.10: Bounds on the performance of binary hard-decision sphere decoding of the binary image of the $(31, 15)$ RS code BPSK modulated on an AWGN channel. The performance of hard-decision sphere decoders with (binary) Hamming radii of $5, 8, 10, 15, 18$ are compared. The bound on the HD-ML decoder is the same for an HD sphere decoder with radius 18. The HD BM and GS symbol based decoders are also compared. The performance of the SD Koetter-Vardy algorithm and the binary SD-ML decoder are plotted for reference.

an BSC channel with a crossover probability p is upper bounded by

$$P(\Sigma_m) \leq \begin{cases} P(\mathcal{E}_{ML}, \zeta < m_o) + P(\zeta \geq m_o), & m \geq m_o \\ P(\mathcal{E}_{ML}, \zeta < m) + P(\zeta \geq m), & m < m_o \end{cases},$$

where the minimizing radius m_o is given by (8.41). $P(\zeta \geq m)$, $P(\mathcal{E}_{ML}, \zeta < m)$ are given by (8.39) and (8.40) respectively.

8.4.3   Numerical Examples

In Figure 8.10, we show bounds on the performance of HD decoding of the near half rate $(31, 15)$ RS code over $\mathbb{F}_{32}$ when its binary image is transmitted over an AWGN channel followed by hard decisions. The optimum binary decoding radius is 18. Thus the closer the decoding radius is to 18, the better the performance of the sphere decoder. The HD-ML decoder has more than 2 dB coding gain over the Berlekamp Massey (BM) decoder, which can correct 8 symbol errors. It is observed that the average performance of an HD sphere decoder, with a (binary Hamming) radius 8, closely upper bounds that of the HD-BM decoder that can correct 8 symbol errors. The HD-GS decoder can correct one more symbol error than the BM decoder. The performance of the GS algorithm is analyzed by modeling it as 16-ary HD sphere decoder of radius 9. Consequently, one can observe that a hard-decision sphere decoder with a binary decoding radius of 10 outperforms the symbol based GS decoder. Surprisingly, the performance of the soft-decision Koetter-Vardy algorithm with infinite interpolation cost almost overlaps with that of a binary hard-decision sphere decoder with radius 15. This might speculate that the performance of the Koetter-Vardy algorithm can be bounded by that of a binary hard-decision sphere decoder with some decoding radius.

In Figure 8.11,the binary image of the $(15, 3)$ RS code is BPSK modulated over an AWGN channel. For 16-ary hard decisions, the channel is modeled as an QSC. The performance bound of the hard ML (H-ML) decoder is shown (Theorem 8.9) and is the same as an HSD of radius 9. The bounds of (8.39) and (8.40) are also shown and labeled as "$F(9)$" and "$E(9)$" respectively. As seen, the three bounds ("bnd") are in close agreement with the simulation ("sim"), for such a hypothetical sphere decoder.

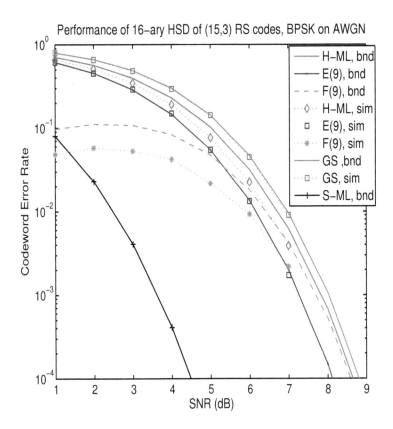

Figure 8.11: The $(15, 3)$ RS code is BPSK modulated and transmitted over an AWGN channel. For the 16-ary hard-decision decoder, the channel is an QSC.

The optimum radius m_o for the ML bound is 9. For the HD-ML decoder, or equivalently a HD sphere decoder with radius 9, the bounds are compared to simulations for a sphere decoding ML error E(9), sphere decoding failure F(9), and their sum H-ML (error plus failure probability) The Guruswami-Sudan (GS) radius is 8 and the corresponding error plus failure probability is plotted. The binary soft-decision ML decoder performance (S-ML) is also plotted.

The error probability of the GS decoder with radius 8 is simulated and agrees with the bound of Theorem 8.11. For reference proposes, we show the average error probability of the soft-decision bit level ML (S-ML) decoder (c.f., [29]) which has about 4 dB gain over the symbol H-ML decoder.

8.5 Complexity of Sphere Decoding

The expected complexity of sphere decoding was thoroughly analyzed in [54]. In Figure 8.12, the empirical complexity exponents of SSD of the $(24, 12)$ Golay code BPSK modulated over an AWGN channel are shown. It is clear that for a larger decoding radius there is a price paid in terms of the complexity. We also show the complexity of the SSD whose radius changes such that with a probability of 0.9 the transmitted word is inside the sphere centered around the received one. In other words, the radius of this sphere is calculated by (see (8.4))

$$r = \arg_D \Gamma_r(n_d/2, D^2/2\sigma^2) = 0.9. \qquad (8.42)$$

The corresponding complexity is labeled "r^2: 90% confidence." As the signal-to-noise ratio increases (σ^2 decreases), this radius decreases. Thus, using this technique, the sphere decoder complexity decreases with the SNR. However, the error plus failure probability will be lower bounded with the failure probability of the sphere decoder (in this case 0.1). At a slighter increase in average complexity one can achieve ML decoding, by starting with the previous radius and gradually increasing the decoding radius until a codeword is found. The corresponding complexity is shown as "r^2: 0.90 + cumulative." For the 90% confidence case, the variation of the radius versus the SNR is shown in Figure 8.13. The radius decreases as the SNR increases as expected from (8.42).

8.6 Conclusion

Bounds on the error plus failure probability of hard-decision and soft-decision sphere decoding of block codes were derived. By comparing with the simulations of the cor-

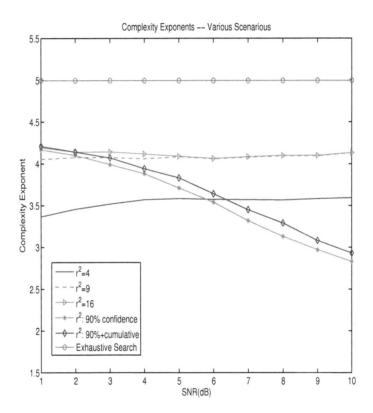

Figure 8.12: Complexity exponent for SD sphere decoding of the $(24, 12)$ Golay code. The complexity exponent (of the number of flops) is plotted versus the SNR for decoders with squared Euclidean radii of 4, 9 and 16 respectively and compared to that of the ML exhaustive-search decoder. The sphere decoder with a failure probability 10 percent is labeled "90% confidence." If the radius of this sphere decoder keeps incremented till a codeword is found, this sphere decoder is labeled "90%+ cumulative."

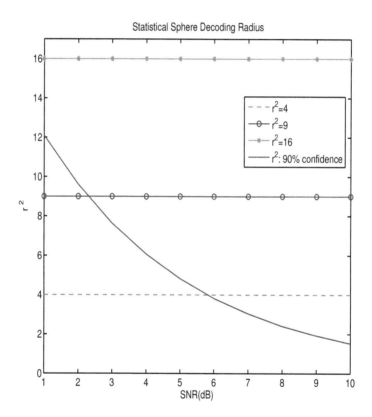

Figure 8.13: Statistical (squared Euclidean) decoding radius versus fixed decoding radius for the $(24, 12)$ Golay code.

responding decoders, we demonstrate that our bounds are tight. The ML performance of codes on q-ary symmetric channels is analyzed. The performance of sphere decoding of Reed-Solomon codes and their binary images was analyzed. Moreover, the bounds are extremely useful in predicting the performance of the sphere decoders at the tail of error probability when simulations are prohibitive. The bounds allows one to pick the radius of the sphere decoder that best fits the performance, throughput and complexity requirements of the system.

Appendix A

Newton's Algorithm

We briefly sketch the Newton algorithm used to minimize an arbitrary function $f(x)$ in m variables. For more details, we refer the reader to [17] and [70]. The gradient of $f(x)$ is the $(m \times 1)$-dimensional vector $\nabla f(x)$, and its $(m \times m)$ Hessian is $H_f(x)$. We assume that $f(x)$ is twice continuously differentiable, there exists at least one solution x_{opt} such that $\nabla f(x_{opt}) = 0$ and the Hessian $H_f(x)$ is positive definite for $x = x_{opt}$.

Let $\mathbf{x_o}$ be the initial iterate, then for iteration n:

1. Test for termination:

Stop if $\|\nabla f(x_n)\| \le \tau_r \|\nabla f(x_o)\| + \tau_a$, τ_r and τ_a are small positive numbers and are called the relative tolerance and absolute tolerance respectively.

2. Find the Newton Direction, d:

Calculate the Hessian, $H_f(x_n)$ if an analytical expression is found, otherwise approximate $H_f(x_n)$ with a finite difference Hessian. The later case involves m new evaluations, $\nabla f(x_n + \delta e_j)$, $j = 1, \ldots, m$ where e_j is the unit vector in the jth coordinate direction. The Newton direction satisfies

$$H_f(x_n)\mathbf{d} = -\nabla f(x_n).$$

This requires the LU factorization of the Hessian using Gaussian elemination, $H_f(x_n) = PLU = L'U$, and solving for $L'z = -\nabla f(x)$ and $U\mathbf{d} = z$. The LU decomposition require $m^3 + O(m^2)$ flops and solving for the triangular systems requires $m^2 + O(m)$ flops. The complexity of the algorithm lies here.

3. Line Search:

The Armijo rule for calculating the length of the Newton step, λ, iteratively finds $\lambda_o, \lambda_1, ...\lambda_k$ till

$$\|\nabla f(\mathbf{x_n} + \lambda_k \mathbf{d})\| < (1 - \alpha\lambda_k)\|\nabla f(\boldsymbol{x_n})\|$$

for the smallest $k \geq 0$ and $\alpha \in (0, 1)$ is typically 10^{-4} to easily satisfy the equation. One method is to let $\lambda_o = 1$ and $\lambda_k = \lambda_{k-1}/2$ for $k \geq 1$. In this implementation, λ_{k+1} is the minimizer of the parabola fitted to the points $\phi(0), \phi(\lambda_k)$ and $\phi(\lambda_{k-1})$ on the interval $[\lambda_k/10, \lambda_k/2]$ where $\phi(\lambda) = \|\nabla f(\mathbf{x} + \lambda\mathbf{d})\|^2$.

4. Update x:

$$\mathbf{x_{n+1}} = \mathbf{x_n} + \lambda\mathbf{d}.$$

Since the Hessian is computationally excessive to compute and factor, a hybrid Chord-Newton strategy is used; the Hessian is updated only after a certain number of nonlinear iterations or if the ratio of successive norms of the nonlinear residuals $\|\nabla f(\boldsymbol{x_n})\|/\|\nabla f(\mathbf{x_{n-1}})\|$ is larger than a certain threshold, i.e., the rate of decrease in the residual is not sufficiently rapid.

Bibliography

[1] P. Agashe, R. Rezaiifar, and P. Bender, "CDMA2000 high rate broadcast packet data air interface design," *IEEE Commun. Magazine*, pp. 83–89, Feb. 2004.

[2] E. Agrell, A. Vardy, and K. Zeger, "Closest point search in lattices," *IEEE Trans. Inform. Theory*, vol. 48, no. 8, pp. 2201–2214, Aug. 2002.

[3] A. Ahmed, R. Koetter, and N. R. Shanbhag, "Performance analysis of the adaptive parity check matrix based soft-decision decoding algorithm," in *Asilomar Conference*, 2004.

[4] S. Aji, H. Jin, A. Khandekar, D. J. Mackay, and R. J. McEliece, "BSC thresholds for code ensembles based on "typical pairs" decoding," in *IMA Workshop on Codes and Graphs*, Aug. 1999, pp. 195–210.

[5] C. Argon, S. McLaughlin, and T. Souvignier, "Iterative application of the Chase algorithm on Reed-Solomon product codes," in *IEEE International Conference on Communications, ICC 2001*, Jun. 2001.

[6] C. Argon and S. W. McLaughlin, "An efficient Chase decoder for turbo product codes," *IEEE Trans. Commun.*, vol. 52, no. 6, pp. 896–898, Jun. 2004.

[7] L. Bahl, J. Cocke, F. Jeinek, and J. Raviv, "Optimal decoding of linear codes for minimizing symbol error rate." *IEEE Trans. Inform. Theory*, vol. 20, pp. 284–287, Mar. 1974.

[8] S. Benedetto, D. Divsalar, G. Montorsi, and F. Pollara, "Serial concatenation of interleaved codes: Performance analysis, design and iterative decoding." *IEEE Trans. Inform. Theory*, vol. 44, no. 3, pp. 909–926, May 1998.

[9] S. Benedetto and G. Montorsi, "Unveiling turbo codes: Some results on parallel concatenated coding schemes," *IEEE Trans. Inform. Theory*, vol. 42, no. 3, pp. 409–428, Mar. 1996.

[10] Berlekamp, R. McEliece, and H. van Tilborg, "On the inherent intractability of certain coding problems," *IEEE Trans. Inform. Theory*, vol. 24, pp. 384–386, May 1978.

[11] E. Berlekamp, "The technology of error-correcting codes," *Proc. IEEE*, vol. 68, no. 8, pp. 564–593, May 1980.

[12] E. R. Berlekamp, *Algebraic Coding Theory*. New York: McGraw-Hill, 1968.

[13] C. Berrou and A. Glavieux, "Near-optimum correcting coding and decoding: Turbo codes," *IEEE Trans. Commun.*, vol. 44, pp. 1261–1271, Oct.

[14] G. Beyer, K. Engdahl, and K. Zigangirov, "Asymptotic analysis and comparison of two coded modulation schemes using PSK signaling-Part I," *IEEE Trans. Inform. Theory*, vol. 47, no. 7, pp. 2782–2792, Nov. 2001.

[15] I. Blake and K. Kith, "On the complete weight enumerator of Reed-Solomon codes." *SIAM J. Disc. Math.*, vol. 4, no. 2, pp. 164–171, May 1991.

[16] M. Blaum, J. Bruck, and A. Vardy, "MDS array codes with independent parity symbols," *IEEE Trans. Inform. Theory*, vol. 42, no. 2, pp. 529–542, 1996.

[17] S. Boyd and L. Vandenberghe, *Convex Optimization*. Cambridge: Cambridge University Press, 2004.

[18] D. Chase, "A class of algorithms for decoding block codes with channel measurement information," *IEEE Trans. Commun.*, vol. 18, pp. 170–182, May 1972.

[19] F. Chiaraluce and R. Garello, "Extended Hamming product codes analytical performance evaluation for low error rate applications," *IEEE Trans. Wireless Commun.*, vol. 3, pp. 2353–2361, Nov. 2004.

[20] D. Coppersmith and M. Sudan, "Reconstructing curves in three (and higher) dimensional space from noisy data," in *STOC'03, San Diego, California, USA.*, Jun. 2003.

[21] M. O. Damen, A. Chkeif, and J. Belfiore, "Lattice code decoder for space-time codes," *IEEE Commun. Lett.*, pp. 161–163, May 2000.

[22] M. O. Damen, H. E. Gamal, and G. Caire, "On maximum-likelihood detection and the search for the closest lattice point," *IEEE Trans. Inform. Theory*, vol. 49, no. 10, pp. 2389–2402, 2003.

[23] D. Divsalar, "A simple tight bound on error probability of block codes with application to turbo codes," TMO Progress Report, NASA, JPL, Tech. Rep. 42–139, 1999.

[24] D. Divsalar and E. Biglieri, "Upper bounds to error probabilities of coded systems over AWGN and fading channels," in *Proc. 2000 IEEE Global Telecommunications Conf. (GLOBECOM00), San Francisco, CA*, Nov. 2000, pp. 1605–1610.

[25] S. Dolinar, D. Divsalar, and F. Pollara, "Code performance as a function of block size," TMO Progress Report, Tech. Rep. 42-133, 1998.

[26] M. El-Khamy, "The average weight enumerator and the maximum-likelihood performance of product codes," in *International Conference on Wireless Networks, Communications and Mobile Computing, WirelessCom Information Theory Symposium, Hawaii*, vol. 2, Jun. 2005, pp. 1587–1592.

[27] M. El-Khamy and R. Garello, "On the weight enumerator and the maximum-likelihood performance of linear product codes," submitted to *IEEE Trans. on Inform. Theory*, Dec. 2005.

[28] M. El-Khamy and R. J. McEliece, "On the multiuser error probability and the maximum-likelihood performance of MDS codes." submitted to *IEEE Trans. on Inform. Theory*, Aug. 2006.

[29] M. El-Khamy and R. J. McEliece, "Bounds on the average binary minimum distance and the maximum-likelihood performance of Reed Solomon codes," in *42nd Allerton Conf. on Communication, Control and Computing*, 2004.

[30] M. El-Khamy and R. J. McEliece, "Iterative algebraic soft-decision decoding of Reed-Solomon codes," in *IEEE International Symposium on Information Theory and its Applications, Parma, Italy*, 2004, pp. 1456–1461.

[31] M. El-Khamy and R. J. McEliece, *Interpolation Multiplicity Assignment Algorithms for Algebraic Soft-Decision Decoding of Reed-Solomon Codes*. DIMACS Series in Discrete Mathematics and Theoretical Computer Science, Algebraic Coding Theory and Information Theory, American Mathematical Society, 2005, vol. 68.

[32] M. El-Khamy and R. J. McEliece, "The partition weight enumerator of MDS codes and its applications." in *IEEE International Symposium on Information Theory, Adelaide, Australia*, Sep. 2005, pp. 926–930.

[33] M. El-Khamy and R. J. McEliece, "Iterative algebraic soft-decision list decoding of Reed-Solomon codes," *IEEE J. Select. Areas Commun.*, vol. 24, no. 3, pp. 481–490, Mar. 2006.

[34] M. El-Khamy, R. J. McEliece, and J. Harel, "Performance enhancements for algebraic soft-decision decoding of Reed-Solomon codes," in *IEEE International Symposium on Information Theory, Chicago, Illinois*, 2004, p. 421.

[35] M. El-Khamy, H. Vikalo, and B. Hassibi, "Bounds on the performance of sphere decoding of linear block codes," in *Proc. of IEEE Information Theory Workshop on Coding and Complexity, ITW2005, Rotorua, New Zealand*, 2005.

[36] M. El-Khamy, H. Vikalo, B. Hassibi, and R. J. McEliece, "Performance of sphere decoding of block codes," submitted to *IEEE Trans. on Commun.*, Feb. 2006.

[37] M. El-Khamy, H. Vikalo, B. Hassibi, and R. J. McEliece, "On the performance of sphere decoding of block codes," in *2006 IEEE International Symposium on Information Theory, Seattle, Washington*, Jun. 2006.

[38] P. Elias, "Error-free coding," *IRE Trans. Inform. Theory*, vol. IT-4, pp. 29–37, Sep. 1954.

[39] P. Elias, "List decoding for noisy channels," MIT Electronics Research Lab, MIT, Tech. Rep. 335, 1957.

[40] U. Fincke and M. Pohst, "Improved methods for calculating vectors of short length in a lattice, including a complexity analysis," *Mathematics of Computation*, vol. 44, pp. 463–471, 1985.

[41] G. D. Forney, "Generalized minimum distance decoding," *IEEE Trans. Inform. Theory*, vol. 12, pp. 125–131, 1966.

[42] M. Fossorier and S. Lin, "Soft-decision decoding of linear block codes based on ordered statistics," *IEEE Trans. Inform. Theory*, vol. 41, pp. 1379–1396, Sep. 1995.

[43] M. Fossorier, S. Lin, and D. Rhee, "Bit-error probability for maximum-likelihood decoding of linear block codes and related soft-decision decoding methods," *IEEE Trans. Inform. Theory*, vol. 44, no. 7, pp. 3083–3090, Nov. 1998.

[44] M. Fossorier, "Critical point for maximum-likelihood decoding of linear block codes," *IEEE Commun. Lett.*, vol. 9, no. 9, 2005.

[45] R. Gallager, *Low Density Parity Check Codes.* MIT: MIT Press, 1963.

[46] W. J. Gross, F. R. Kschischang, R. Kötter, and P. G. GulakR, "Towards a VLSI architecture for interpolation-based soft-decision Reed-Solomon decoders," submitted to the Journal of VLSI Signal Processing Special Issue on SIPS, preprint dated July 1, 2003.

[47] P. Gupta and P. R. Kumar, "The capacity of wireless networks," *IEEE Trans. Inform. Theory*, vol. 46, no. 2, pp. 388–404, 2000.

[48] V. Guruswami and A. Rudra, "Explicit capacity-achieving list-decodable codes," in *Electronic Colloquium on Computational Complexity (ECCC) Tech Report TR05-133. Nov. 2005.*

[49] V. Guruswami and M. Sudan, "Improved decoding of Reed-Solomon codes and algebraic geometry codes," *IEEE Trans. Inform. Theory*, vol. 45, no. 6, pp. 1757–1767, Sep. 1999.

[50] V. Guruswami and A. Vardy, "Maximum-likelihood decoding of Reed Solomon codes is NP-hard," 2006, submitted to *IEEE Trans. Inform. Theory.*

[51] J. Hagenauer and P. Hoher, "A Viterbi algorithm with soft-decision outputs and its applications," in *GLOBECOM'89, Dallas, Texas*, 1989, pp. 47.1.1–47.1.7.

[52] J. Hagenauer, E. Offer, and L. Papke, "Iterative decoding of binary block and convolutional codes," *IEEE Trans. Inform. Theory*, vol. 42, pp. 429–449, Mar. 1996.

[53] T. Halford, V. Ponnampalam, A. Grant, and K. Chugg, "Soft-in soft-out decoding of Reed-Solomon codes based on Vardy and Be'ery's decomposition," *IEEE Trans. Inform. Theory*, vol. 51, no. 12, pp. 4363–4368, Dec. 2005.

[54] B. Hassibi and H. Vikalo, "On the sphere-decoding algorithm: I. Expected complexity," *IEEE Trans. Signal Processing*, vol. 53, pp. 2806–2818, Aug. 2005.

[55] H. Herzberg and G. Poltyrev, "Techniques of bounding the probability of decoding error for block coded modulation structures," *IEEE Trans. Inform. Theory*, pp. 903–911, May 1994.

[56] H. Herzberg and G. Poltyrev, "The error probability of M-ary PSK block coded modulation schemes," *IEEE Trans. Commun.*, vol. 44, no. 4, pp. 427–433, Apr. 1996.

[57] S. A. Hirst, B. Honary, and G. Markarian, "Fast chase algorithm with an application in turbo decoding," *IEEE Trans. Commun.*, pp. 1693–1699, Oct. 2001.

[58] Hochwald and S. ten Brink, "Achieving near-capacity on a multiple-antenna channel," *IEEE Trans. Commun.*, vol. 53, pp. 389–399, Mar. 2003.

[59] R. Horn and C. Johnson, *Matrix Analysis.* Cambridge: Cambridge University Press, 1985.

[60] T.-H. Hu and S. Lin, "An efficient hybrid decoding algorithm for Reed-Solomon codes based on bit reliability," *IEEE Trans. Commun.*, vol. 51, no. 7, pp. 1073–1081, July 2003.

[61] B. Hughes, "On the error probability of signals in additive white Gaussian noise," *IEEE Trans. Inform. Theory*, pp. 151–155, Jan. 1991.

[62] T.-Y. Hwang, "A relation between the row weight and column weight distributions of a matrix," *IEEE Trans. Inform. Theory*, vol. 27, pp. 256–257, Mar. 1981.

[63] J. Jiang and K. Narayanan, "Iterative soft-decision decoding of Reed-Solomon codes," *IEEE Commun. Lett.*, vol. 8, pp. 244–246, Apr. 2004.

[64] J. Jiang and K. Narayanan, "Iterative soft-decision decoding of Reed Solomon codes based on adaptive parity check matrices," in *Proc. ISIT*, 2004.

[65] J. Jiang and K. R. Narayanan, "Iterative soft-input soft-output decoding of Reed-Solomon codes by adapting the parity-check matrix," *IEEE Trans. Inform. Theory*, vol. 52, no. 8, pp. 3746–3756, Aug. 2006.

[66] M. Kan, Sony Corp., private communication.

[67] T. Kasami and S. Lin, "The binary weight distribution of the extended $(2^m, 2^m - 4)$ code of the Reed-Solomon code over GF(2^m) with generator polynomial $(x - \alpha)(x - \alpha^2)(x - \alpha^3)$," *Linear Algebra Appl.*, pp. 291–307, 1988.

[68] T. Kasami, S. Lin, and W. Peterson, "New generalizations of the Reed-Muller codes–I: Primitive codes," *IEEE Trans. Inform. Theory*, pp. 189–199, Mar. 1968.

[69] T. Kasami, T. Takata, K. Yamachita, T. Fujiwara, and S. Lin, "On bit error probability of a concatenated coding scheme," *IEEE Trans. Commun.*, vol. 45, no. 5, pp. 536–543, May 1997.

[70] C. T. Kelley, *Solving Nonlinear Equations with Newton's Method.* Society for Industrial and Applied Mathematics, Philadelphia, 2003.

[71] R. Koetter, "On algebraic decoding of algebraic geometric and cyclic codes," *Ph.D. thesis, University of Linköping, Sweden*, 1996.

[72] R. Koetter and A. Vardy, "Algebraic soft-decision decoding of Reed-Solomon codes," *IEEE Trans. Inform. Theory*, vol. 49, no. 11, pp. 2809–2825, Nov. 2003.

[73] H. Lu, P. V. Kumar, and E. Yang, "On the input-output weight enumerators of product accumulate codes," *IEEE Commun. Lett.*, vol. 8, no. 8, Aug. 2004.

[74] F. J. MacWilliams and N. J. Sloane, *The Theory of Error Correcting Codes.* Amsterdam: North Holland, 1977.

[75] R. J. McEliece, *The Theory of Information and Coding*, 2nd ed. Cambridge: Cambridge University Press, 2002.

[76] R. J. McEliece, "The Guruswami-Sudan decoding algorithm for Reed-Solomon codes," IPN Progress Report, Tech. Rep. 42–153, May 15 2003.

[77] R. J. McEliece, "On the average list size for the Guruswami-Sudan decoder," in *ISCTA03*, 2003.

[78] R. J. McEliece, D. MacKay, and J. Cheng, "Turbo decoding as an instance of Pearl's belief-propagation algorithm," *IEEE J. Select. Areas Commun.*, vol. 16, pp. 140–152, Feb. 1998.

[79] R. J. McEliece and L. Swanson, "On the decoder error probability of Reed-Solomon codes," *IEEE Trans. Inform. Theory*, vol. 32, no. 5, pp. 701–703, Sep. 1986.

[80] R. Nielsen and T. Hoeholdt, "Decoding Reed-Solomon codes beyond half the minimum distance," in *Cryptography and Related Areas*, J. Buchmann, T. Hoeholdt, H. Stichenoth, and H. Tapia-Recillas, Eds. Springer-Verlag, 2000, pp. 221–236.

[81] F. Parvaresh and A. Vardy, "Correcting errors beyond the Guruswami-Sudan radius in polynomial time," in *FOCS, 2005*.

[82] F. Parvaresh and A. Vardy, "Multivariate interpolation decoding beyond the Guruswami-Sudan radius," in *Proc. 42nd Annual Allerton Conference on Communication, Control and Computing, Urbana, IL., Oct. 2004*.

[83] F. Parvaresh and A. Vardy, "Multiplicity assignments for algebraic soft-decoding of Reed-Solomon codes," in *IEEE International Symposium on Information Theory*, 2003.

[84] F. Parvaresh, M. El-Khamy, M. Stepanov, D. Augot, R. J. McEliece, and A. Vardy, "Algebraic list decoding of Reed-Solomon product codes," in *Tenth International Workshop on Algebraic and Combinatorial Coding Theory (ACCT-10) Zvenigorod, Russia*, Sep. 2006.

[85] L. Pecquet, "List decoding of algebraic-geometric codes," *PhD thesis, University of Paris*, 2001.

[86] R. Pellikaan and X.-W. Wu, "List decoding of q-ary Reed-Muller codes," *IEEE Trans. Inform. Theory*, pp. 679 – 682, Apr. 2004.

[87] G. Poltyrev, "Bounds on the decoding error probability of binary linear codes via their spectra," *IEEE Trans. Inform. Theory*, vol. 40, no. 4, pp. 1284–1292, July 1994.

[88] V. Ponnampalam and B. Vucetic, "Soft-decision decoding of Reed-Solomon codes," *IEEE Trans. Commun.*, vol. 50, pp. 1758–1768, Nov. 2002.

[89] J. G. Proakis, *Digital Communications*, 4th ed. New York: McGraw-Hill, 2001.

[90] R. Pyndiah, "Near optimum decoding of product codes: Block turbo codes," *IEEE Trans. Commun.*, vol. 46, no. 8, pp. 1003–1010, Aug. 1998.

[91] R. Pyndiah, A. Glavieux, A. Picart, and S. Jacq, "Near optimum decoding of product codes," in *Proc. of IEEE GLOBECOM Conf.*, 1994.

[92] N. Ratnakar and R. Koetter, "Exponential error bounds for algebraic soft-decision decoding of Reed-Solomon codes," *IEEE Trans. Inform. Theory*, vol. 15, no. 11, pp. 3899–3917, Nov. 2005.

[93] I. S. Reed and G. Solomon, "Polynomial codes over certain finite fields," *J. Soc. Industrial Appl. Math*, vol. 8, pp. 300–304, 1960.

[94] C. Retter, "The average binary weight enumerator for a class of generalized Reed-Solomon codes," *IEEE Trans. Inform. Theory*, vol. 37, no. 2, pp. 346–349, Mar. 1991.

[95] R. Roth and G. Ruckenstein, "Efficient decoding of reed-solomon codes beyond half the minimum distance," *IEEE Trans. Inform. Theory*, vol. 46, no. 1, pp. 246–257, 2000.

[96] I. Sason and S. Shamai, "Bounds on the error probability for block and turbo-block codes," *Annals of Telecommunications*, vol. 54, no. 3.

[97] I. Sason and S. Shamai, "Performance analysis of linear codes under maximum-likelihood decoding: A tutorial," *Foundations and Trends in Communications and Information Theory*, vol. 3, July 2006.

[98] I. Sason and S. Shamai, "Improved upper bounds on the ML decoding error probability of parallel and serial concatenated turbo codes via their ensemble distance spectrum," *IEEE Trans. Inform. Theory*, vol. 46, no. 1, pp. 24–47, Jan. 2000.

[99] I. Sason, S. Shamai, and D. Divsalar, "Tight exponential upper bounds on the ML decoding error probability of block codes over fully interleaved fading channels," *IEEE Trans. Commun.*, vol. 51, no. 8, pp. 1296–1305, Aug. 2003.

[100] C. Schnorr and M. Euchner, "Lattice basis reduction: Improved practical algorithms and solving subset sum problems," *Math. Programming*, vol. 66, pp. 181–191, 1994.

[101] C. E. Shannon, "Probability of error for optimal codes in a Gaussian channel," *Bell Syst. Tech. J.*, vol. 38, pp. 611–656, 1959.

[102] M. Sudan, "Decoding of Reed-Solomon codes beyond the error-corrrection bound," *J. Complexity*, vol. 13, pp. 180–193, 1997.

[103] H. Tang, Y. Liu, M. Fossorier, and S. Lin, "On combining Chase-2 and GMD decoding algorithms for nonbinary block codes," *IEEE Commun. Lett.*, vol. 5, no. 5, pp. 209–211, May 2001.

[104] M. Tanner, "A recursive approach to low complexity codes," *IEEE Trans. Inform. Theory*, vol. 27, no. 5, pp. 533–547, 1981.

[105] L. Tolhuizen, "More results on the weight enumerator of product codes," *IEEE Trans. Inform. Theory*, vol. 48, no. 9, pp. 2573–2577, Sep. 2002.

[106] L. Toluizen, S. Baggen, and E. Hekstra-Nowacka, "Union bounds on the performance of product codes," in *Proc. of ISIT 1998.*, Cambridge, MA, USA, 1998.

[107] D. Torrieri, "Information-bit, information-symbol, and decoded-symbol error rates for linear block codes," *IEEE Trans. Commun.*, pp. 613–617, May 1988.

[108] A. Valembois and M. Fossorier, "Sphere-packing bounds revisited for moderate block lengths," *IEEE Trans. Inform. Theory*, vol. 50, no. 12, pp. 2998–3014, Dec. 2004.

[109] J. H. van Lint and R. M. Wilson, *A Course in Combinatorics*, 2nd ed. Cambridge: Cambridge University Press, 2001.

[110] H. Vikalo and B. Hassibi, "On joint detection and decoding of linear block codes on Gaussian vector channels," to appear in *IEEE Trans. on Signal Processing*.

[111] H. Vikalo and B. Hassibi, "Statistical approach to ML decoding of linear block codes on symmetric channels," in *Proceedings of IEEE International Symposium on Information Theory (ISIT)*, 2004.

[112] A. Viterbi, "Error bounds on convolutional codes and an assymptotically optimum decoding algorithm," *IEEE Trans. Inform. Theory*, vol. 13, pp. 260–269, Apr. 1967.

[113] E. Viterbo and J. Boutros, "A universal lattice decoder for fading channels," *IEEE Trans. Inform. Theory*, vol. 45, p. 1639.

[114] E. W. Weisstein, *Mathworld–A Wolfram Web Resource.* http://mathworld.wolfram.com.

[115] S. B. Wicker, *Error Control Systems for Digital Communication and Storage.* Prentice Hall, 1995.

[116] S. B. Wicker and M. J. Bartz, "Type-II hybrid- ARQ protocols using punctured MDS codes," *IEEE Trans. Commun.*, vol. 42, pp. 1431–1440, Feb./Mar./Apr. 1994.

[117] J. M. Wozencraft, "List decoding," Quart. Progress Report, Research Lab. Electronics, MIT, Tech. Rep. 48, 1958.

[118] J. M. Wozencraft and I. M. Jacobs, *Principles of Communication Engineering.* John Wiley & Sons, Inc., 1965.

[119] X.-W. Wu, "An algorithm for finding the roots of the polynomials over order domains," in *Proc. of IEEE International Symposium on Information Theory, Lausane, Switezerland*, Jun. 2002, p. 202.

[120] X.-W. Wu and P. H. Siegel, "Efficient root-finding algorithm with application to list decoding of algebraic-geometric codes," *IEEE Trans. Inform. Theory*, vol. 47, no. 6, pp. 2579–2587, Sep. 2001.

[121] J. Yedidia, W. Freeman, and Y. Weiss, "Understanding belief-propagation and its generalizations," *Exploring Artificial Intelligence in the New Millennium, ISBN 1558608117*, pp. 239–236, Jan. 2003.

[122] R. W. Yeung, S.-Y. R. Li, N. Cai, and Z. Zhang, "Network coding theory: Single sources," *Foundations and Trends in Communications and Information Theory*, vol. 2, Jun. 2005.